EARLY LIFE

A SERIES OF BOOKS IN BIOLOGY
CONSULTING EDITOR: Cedric I. Davern

LYNN MARGULIS
Boston University

Early Life

Science Books International, Incorporated
Boston, Massachusetts

COVER: *Marine*, an oil painting by Gustave Courbet (1819–1877) in the Norton Simon Museum, Pasadena, California, shows how the world might have looked about two billion years ago. Reproduced by permission of The Norton Simon Foundation, Pasadena.

Science Books International, Publishers, is a division of Science Books International, Inc., with editorial offices at 30 Granada Court, Portola Valley, California 94025.

Offices for sales and customer service are at 51 Sleeper Street, Boston, Massachusetts 02210.

Library of Congress Cataloging in Publication Data

Margulis, Lynn
Early life.

Includes bibliographies, glossary, and index.
1. Life—Origin. 2. Life (Biology) I. Title.
QH325.M29 577 81-18337
Hardcover ISBN 0-86720-003-0 AACR2
Paperback ISBN 0-86720-005-7

STAFF FOR THIS BOOK: *Publisher:* Arthur C. Bartlett. *Manuscript Editor:* Andrew Kudlacik. *Book and Cover Design:* Elizabeth W. Thomson. *Illustrators:* Laszlo Meszoly and Julia Gecha. *Production:* Michael Michaud and Elizabeth W. Thomson, Unicorn Production Services, Inc. *Composition:* Palatino, set by Achorn Graphic Services, Inc. *Color Printing:* John P. Pow Company. *Printing and Binding:* Halliday Lithograph Corporation.

Printed in the United States of America

9 8 7 6 5 4 3 2 1

To my children:
Dorion, Jeremy, Zachary, and Jenny

Preface

For most of the history of life on this planet, the living landscape resembled Gustave Courbet's time-forgotten seashore reproduced on the cover of this book. Although inconspicuous, life in the form of bacteria and their diverse communities changed forever the surface and atmosphere of the planet. Although tiny, early life was complex and original. In mudflats, evaporite expanses, fens, and ponds, microbes evolved innovations that we now associate with animals and plants: reproduction, predation, movement, self-defense, sexuality, and many others. This book attempts to tell these stories of early life. I hope it conveys some of the excitement in the current attempts to reconstruct the opening chapters of life on the planet Earth, long before the appearance of the simplest animal or plant.

Are the well-formed filaments found so recently in the Warrawoona Series of northwestern Australia really evidence of the oldest life on the planet? Do the fossils found in the great Gunflint Iron Formation of Ontario tell us that bacteria were instrumental in the accumulation of the most important iron reserves in the world? These questions are not solved here, but they are raised for students, scientists, and general readers interested in earliest evolution and its consequences. No special scientific background is required of the reader, only a lively interest.

This book was inspired by Gerard Piel and Edward Immergut of *Scientific American.* I am indebted to their inspiration and especially to Toni Gerber who spent long and productive hours at *Scientific American* with the manuscript in its early stages. The largest debt of all is to Andrew Kudlacik, who performed all editing tasks with zeal and competence. In the making of this book, he became a serious student of microbiology. It was he who translated arcane metabolic schemes into clear-breathing prose and diagrams. I am also grateful to Laurie Read, Michael Michaud, Lydia Stiver, Susan Lenk, Dorion Sagan, and Elizabeth Thomson for aid in preparing the manuscript and to my publisher, Arthur Bartlett, for his unfailing encouragement. Laszlo Meszoly, Julia Gecha, and Linda Reeves have graced the pages with fine art work. Without Jeremy Sagan's computer program there would be no index. Many students and colleagues provided information and illustrations—Professors E. S. Barghoorn and S. W. Awramik deserve special mention. I am grateful to Dr. Cedric I. Davern for comments on the manuscript.

Some of the research described in this book was supported by the Planetary Biology program of the National Aeronautics and Space Administration and some by the Guggenheim Foundation (fellowship, 1979). We are grateful to NASA and to the Boston University Graduate School for the continuing opportunity to do research.

Lynn Margulis

Contents

Illustrations

Illustration Acknowledgments

Figures 1-1, 2-16, 3-8, and 4-14 are redrawn from *Symbiosis in Cell Evolution: Life and its Environment on the Early Earth* by Lynn Margulis. W. H. Freeman and Company. Copyright © 1981.

Figures 4-9 (right) and 6-2 are from, and Figure 1-6 is redrawn from, *Five Kingdoms: An Illustrated Guide to the Phyla of Life on Earth* by Lynn Margulis and Karlene V. Schwartz. W. H. Freeman and Company. Copyright © 1982.

Figure 1-2. Copyright © 1975 by the American Association for the Advancement of Science (from Wireman, J. W. and Dworkin, M. "Morphogenesis and Developmental Interactions in Myxobacteria," *SCIENCE*, Vol. 189, 14 August 1975, pp. 516–523).

Figure 3-2 reprinted by permission from Nature (5526), pp. 489–492. Copyright © 1975, Macmillan Journals Limited.

Figure 4-1 (top) reprinted from *Current Topics in Cellular Regulation*, Vol. 2, A. W. Linnane, "The Biogenesis of Mitochondria," pp. 101–172. New York: Academic Press, Inc., 1970.

Figure 4-1 (bottom) reprinted by permission from the *Journal of Experimental Botany*, Vol. 32 No. 127 (1981), pp. 311–320.

Figure 4-9 (left) reprinted by permission from Daniels, E. W., *The Biology of the Amoeba*, Chapter 5 "Ultrastructure." New York: Academic Press, Inc., 1972.

Figure 4-16 reprinted by permission from Paulin, J. J. and Bussey, J., "Oral regeneration in the ciliate *Stentor coeruleus:* a scanning and transmission electron optical study," *The Journal of Protozoology*, Vol. 18, pp. 201–213, 1971.

EARLY LIFE

CHAPTER 1

Evolution and Cells

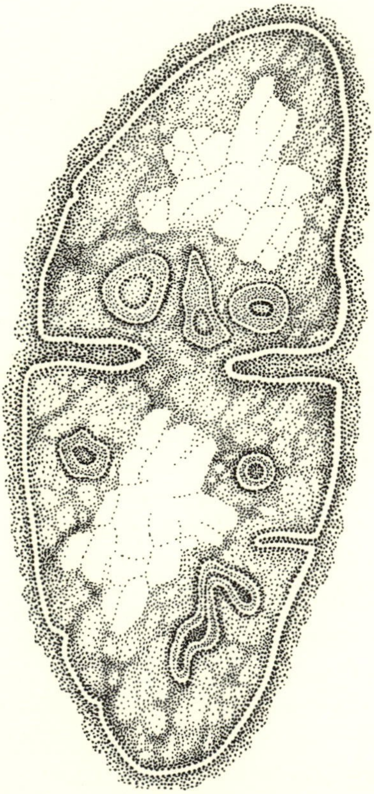

THE EARTH has had a solid surface of rocks for about four billion years. The oldest fossils—of microscopic isolated spheres resembling modern bacteria—are about 3.5 billion years old. Yet until about half a billion years ago, no large multicelled organisms—no animals or plants—inhabited the Earth. At about that time, the fossil record shows, marine animals appeared all along the world's seashores. From these animals and the seaweeds that fed them have descended many forms of life. Since then, life has crawled out on the land, flowering plants have appeared and become the dominant vegetation, and all the insects, fishes, reptiles, birds, and mammals have appeared. The history of human beings is a mere moment compared with what went before—the first biologically modern human remains, of *Homo sapiens,* appear in the fossil record of only about 30,000 years ago.

Is evolution going faster and faster? Why did it take three billion years for the elaboration of the single cell into the large multicelled organism? The story of this prolonged interval in evolution is the theme of this book. It is an account of the evolution of early cells. These organisms invented the chemical and biological strategies that made more intricate life forms possible. During those first three billion years, the cell went through profound evolutionary development; it was engaged, quite literally, in evolving its working parts. By the time marine algae and animals appeared, microbes had developed all the major biological adaptations: diverse energy-transforming and feeding strategies, movement, sensing, sex, and even cooperation and competition. They had invented nearly everything in the modern repertoire of life except, perhaps, language and war.

Until recently, most efforts at reconstructing the ways in

which organisms have evolved were directed toward animals and plants. That the simpler but more abundant and diverse microorganisms are also products of a long evolutionary history is a new realization, one that has developed from recent discoveries in several fields, including microbiology, biochemistry, and geology. Perhaps the most illuminating discoveries have been made by use of the electron microscope, which, using an electron beam instead of light, can magnify as much as 500,000 times. Organisms thought to be similar have turned out to be full of surprising differences; structures and organisms apparently quite diverse have turned out to have a great deal in common. Knowledge of the detailed fine structure of cells has led directly to insight into evolutionary relationships.

Much evidence for evolutionary history is gained by studying extant organisms. Fortunately for students of ancient life, successful innovation perpetuates itself; once complicated patterns of growth and metabolism arise and thrive, they tend to persist. The minuscule bacteria have become optimally adapted to such ancient and persistent niches as rocky seashores, mud flats, stream beds, and salt flats. By studying patterns of metabolism, gas exchange, and behavior in these ubiquitous cells, researchers have begun to piece together a picture of what their earliest ancestors were like.

Two Kinds of Life

The cells of large organisms, such as plants and animals, generally are larger than bacterial cells. They also differ in other fundamental ways. Animal and plant cells always contain organelles, distinct intracellular organs that differ recognizably from their surroundings in the cell. One organelle that they all have is the nucleus. Separated from the rest of the cell by a membrane, the nucleus is a bag that contains the genetic material, deoxyribonucleic acid (DNA), as well as crucial large protein molecules and ribonucleic acid (RNA). By definition, a cell that contains its DNA in a membrane-bounded nucleus is a eukaryote. The living world is unambiguously divisible into eukaryotes and prokaryotes, cells that lack nuclei. All large and elaborate forms of life are composed of eukaryotic ("truly nucleated") cells, whereas bacteria and their microbial relatives are composed of prokaryotic ("pre-nucleated") cells (see Figure 1-1).

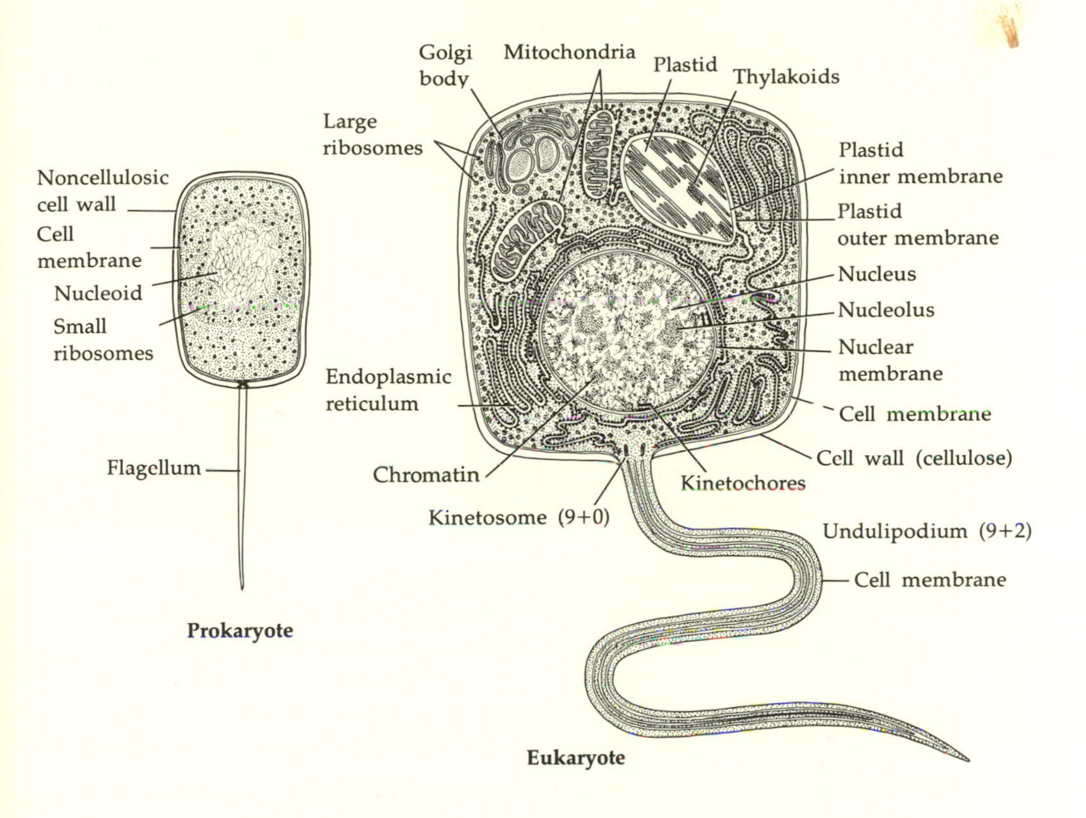

Figure 1-1. Prokaryotic and eukaryotic cells compared, as they would appear under a transmission electron microscope. Most of these organelles and other structures will be described in the following chapters. Not every eukaryotic cell exhibits all these features. For example, animal and fungal cells do not have plastids, which are in all photosynthesizing organisms. Few plant cells and no fungal cells have undulipodia.

Chromatin is a complex of DNA and protein diffused throughout the cell nucleus most of the time. As a eukaryotic cell prepares to divide, the chromatin condenses into rod-shaped bodies—the chromosomes—and in many the nucleolus disappears.

In eukaryotes, the endoplasmic reticulum is an intricate network of membranes that extends through much of the cytoplasm, the part of the cell outside the nucleus. It connects the nuclear membrane with the outer membrane of plastids and mitochondria. Much of it is covered with ribosomes, the small bodies that translate genetic instructions from the nucleus into specific proteins (see Figure 1-4). [Drawings by Laszlo Meszoly.]

In eukaryotic cells, the DNA is tightly coiled with protein into chromosomes, rodlike bodies inside the nucleus. The DNA of prokaryotes, in contrast, is a single long circular molecule of DNA that floats free in the interior of the cell. With very few exceptions, all eukaryotic cells contain mitochondria, membrane-bounded organelles in which oxygen is used to "burn" food molecules that provide energy for most other cell activities. Another organelle that generates energy is the chloroplast, a membrane-enclosed unit that contains chlorophyll. Cells of green plants and green algae contain at least one and as many as hundreds or even millions of chloroplasts. They are the sites of photosynthesis, the process by which cells transform the energy of sunlight into chemical energy. In prokaryotes, the consumption of food molecules and the process of photosynthesis are not confined to enclosed organelles, but take place on membranes distributed widely through the cell.

Motile eukaryotic cells typically carry on their external membranes short hairlike structures (cilia) or longer whiplike structures (flagella). Both cilia and flagella* are made of bundles of small hollow microtubules arranged in an elaborate pattern. The beating of these hairs and whips can move the cell itself or can move particles and fluids past a stationary cell. Among prokaryotes, the analogous structures (also called flagella) are far smaller and simpler—they are single stranded. Other components unique to eukaryotic cells are centrioles, small, dotlike bodies that appear during cell division; vacuoles, membrane-enclosed spaces that take part in fluid and salt regulation; lysosomes, small packages of chemicals that break down food particles for intracellular digestion; and Golgi bodies, groups of flattened membranous sacs that package and transport products synthesized by specialized cells. Golgi bodies are especially conspicuous in cells that produce hard shells, skeletons, or glandular secretions such as digestive juices.

The earliest life on Earth consisted only of simple prokaryotic cells. Organisms made of eukaryotic cells did not appear on the scene until much later. Precisely when this evolutionary innovation took place has been the subject of much debate. Eukaryotic

*Called undulipodia (see p. 97).

cells may be more than two billion years old, but they cannot be less than about 700 million years old; by that time, marine animals made of such cells were distributed along many seashores. How did eukaryotic cells arise? The sequence of events linking prokaryotic ancestors with their eukaryotic descendants is the subject of wide discussion, and different hypotheses—the subjects of many laboratory investigations—have been put forward. The theory I favor is that certain organelles of the eukaryotic cells originated by symbiosis.

Symbiosis can be defined as the intimate living together of two or more organisms, called symbionts, of different species. According to the symbiotic theory of the origin of eukaryotes, once-independent microbes came together, first casually as separate guest and host cells, then by necessity. Eventually, the guest cells become the organelles of a new kind of cell. Such a sequence of events can be found in the symbiotic relationships between many modern life forms. Many organisms live inside, on top of, or attached to other organisms. Hereditary symbioses—those in which the partners remain together throughout their life cycle—are surprisingly common. In some instances, one partner can manufacture its own food by photosynthesis, but the other cannot. Organisms of the first type, able to capture the Sun's energy directly and use it to synthesize the compounds they need for growth and reproduction, are known as autotrophs ("fed by self," from Greek *trophe*, nourishment). Organisms of the second type are called heterotrophs ("fed by others").

Lichens are a common example of a symbiotic relationship. Characteristically flat, crusty, plantlike organisms that can survive in alternately wet and dry and harshly cold environments, lichens are symbiotic partnerships between algae (autotrophs) and fungi (heterotrophs). The algal cells are enfolded by tough threadlike fungal cells, which protect the algae from the harshness of their environment. The algae, which must live in water when they live independently, produce food photosynthetically for themselves and for their fungal partners.

Certain bacteria that inhabit the mud of lake bottoms also enter into symbioses. A larger partner that can swim will team up with several smaller, immobile forms that can produce their own food by photosynthesis. This consortium of bacteria then swims as a unit which has the capabilities of both partners.* Coral reefs, too, depend on the association between tiny coelenterates (corals, jellyfish, sea anemones) and their symbiotic partners, typically single-celled dinoflagellates of the genus *symbiodinium*. The dinoflagellates, which live inside the cells of their hosts, photosynthesize food that supports thriving populations of reef dwellers in nutrient-poor waters.

Nearly every group of organisms has members that have formed close partnerships for feeding, cleaning, or protection. The physiology and patterns of inheritance of modern symbionts provide analogies for evaluating the hypothesis that cell organelles arose through symbiosis. I shall explore this hypothesis in detail in Chapter 4.

Kingdoms of Organisms

Traditionally, biologists have placed all organisms in either the animal or the plant kingdom, a division based on the most conspicuous difference between living things: whether they move and search for food or stay put and derive food directly from sunlight. Of course, animals belong to the first category and plants to the second. However, at least since the middle of the nineteenth century, organisms have been known that are difficult to classify as either plant or animal. Euglenas, for example, are microscopic motile organisms that usually inhabit fresh water. Because they can swim and swallow solid food, one would be inclined to place them in the animal group. Yet they also contain bright green chloroplasts and make food by photosynthesis, as plants do. Generations of biologists have been troubled by the need to force such organisms into the plant or the animal kingdom.

Occasionally, solutions to this problem have been proposed—extra kingdoms for problematic organisms. Until recently, such proposals were generally considered—if they were considered at all—as the special pleading of eccentrics. Only in the past decade or so have biologists generally realized that the

*The partners have not been identified. The consortium, like lichens, is named as if it were a single organism. It is called *Pelochromatium*.

fundamental division in the living world is not between plants and animals, but between prokaryotes and eukaryotes. This realization has tended to make multi-kingdom schemes more respectable. One five-kingdom system has gathered support ever since it was proposed, in 1959, by the late R. H. Whittaker, of Cornell University (see Table 1-1).

Table 1-1. The Classification of Some Organisms in the Traditional System and in the Whittaker Five-Kingdom System

	Traditional	Whittaker
Escherichia coli (gut bacterium)	Plant	Moneran
Euglena	Plant	Protoctist
Macrocystis pyrifera (giant kelp)	Plant	Protoctist
Saccharomyces cerevisiae (brewer's yeast)	Plant	Fungus
Allium sativum (garlic)	Plant	Plant
Anopheles quadrimaculatus (malaria mosquito)	Animal	Animal

In his scheme, Kingdom Monera contains all the prokaryotes—bacteria and their relatives. Although many members of this kingdom are strikingly complicated, all are microbial and usually a high-power microscope is needed to see their details. The remaining four kingdoms are all eukaryotic: protoctists, plants, animals, and fungi.* The protoctists comprise all the single-celled eukaryotic organisms (called protists) and their close multicellular relatives. Some protoctists are plantlike—for example, euglenas, dinoflagellates (single-celled organisms, some species of which cause red tides), and brown algae such as giant kelp. Other protoctists form extensive jellylike coats over rotting logs (certain slime molds) or scum on fish (water molds). The majority of ciliates are free-living aquatic forms, but some

*We use a somewhat modified Whittaker five-kingdom scheme in this book.

can live in the rumen stomachs of bovine animals, where they may be involved in grass digestion.

It is commonly thought that multicellularity is a feature only of large eukaryotic organisms. In fact, multicellular organisms can be found in all five kingdoms, even in Monera (see Figure 1-2). What distinguishes the multicellular systems of animals and plants from those of their microbial ancestors is the sophistication of the connections between their cells. Structurally elaborate cell junctions prevail, especially in animals. Apparently, formation of such connections was a prerequisite to cell differentiation—the tissue, organ, and organ-system development that characterizes large organisms.

Living Clues to the Past

Even when there is little direct evidence from the fossil record, there are other clues to ancient history. For example, because nearly all insects have six legs and hard outer skeletons, it is

Figure 1-2. A fruiting body of the multicellular myxobacterium *Chondromyces apiculatus.* In forming such structures, the myxobacteria are unique among the prokaryotes. Myxobacteria in the vegetative stage live as swarms of single cells. These particular cells glide on solid surfaces in contact with each other. They feed on other bacteria and reproduce by cell division. When food becomes scarce, the cells in a swarm migrate together into several mounds. Each mound sends up a stalk topped by a cluster of capsules containing numbers of resting cells. When the capsules break, the cells are scattered by air currents, some to better feeding grounds, where the cycle begins again. In some species, the fruiting structure contains more than a billion cells and can be seen with a hand lens. [Courtesy of Martin Dworkin, University of Minnesota, Minneapolis.]

likely that the common ancestors of all insects also had such features. Similarly, because dogs, cows, giraffes, horses, and even human beings have four limbs with five fingers or toes on each paw or hand, four-legged, five-toed common ancestors can be inferred for these mammals. By similar inference, can one reconstruct the common ancestor of all forms of life on Earth? Certain inherited features peculiar to certain groups, such as hard outer skeletons or four-leggedness, are considered derivatives. They are not required for all forms of life, but are products of evolution that appeared after the origin of life itself. Truly universal features, however, must be the most primitive.

Universal features are found at the biochemical level. All cells generate energy either from sunlight or from energy-rich food molecules. They all produce the energy-rich molecule adenosine triphosphate (ATP), used to power energy-requiring reactions. All contain replicating systems based on the molecules DNA and RNA, without which reproduction and growth would be impossible. The universality of these molecules in all organisms living today tells us that all life on Earth is ultimately related. Of course, present-day life on Earth might have ancestors with molecular features very different from those that organisms now possess, and many such forms may have become extinct before the origin of the recognizable line of organisms to which all life belongs. However, if there ever was a kind of life built on chemical principles different from those that govern all present life, it has died out without a trace. Therefore, one can trace life back only to ancestors that contained the replicating machinery based on the DNA/RNA chemistry found in all modern animals, plants, and other known life forms.

The accurate replication of DNA molecules is crucial for the reproduction of organisms. It ensures that elephants give birth only to elephants and that maple trees give rise only to more maple trees. DNA replication, modified by mutation and natural selection, also explains how elephants and maple trees now live on an Earth that was once populated only by microbes. The importance of DNA in reproduction and also in the growth of cells and organisms lies in the information contained in genes, linear sections of a DNA molecule. Genes carry the information required for the cells faithfully to reproduce themselves.

DNA molecules are polymers, long chains of similar units called nucleotides. Each nucleotide is made of a phosphate group, a sugar molecule (deoxyribose), and a nitrogen-containing organic base (see Figure 1-3). DNA molecules may be very large, containing millions of nucleotides. However, there are only four different nucleotides, and they differ only in their organic bases: adenine, guanine, cytosine, and thymine. RNA molecules, gen-

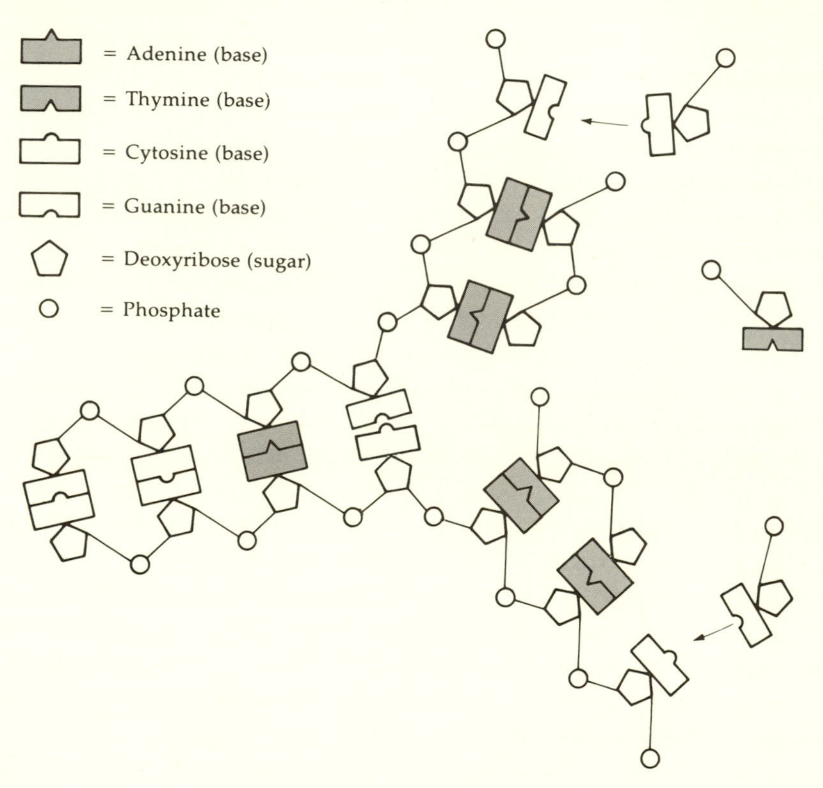

Figure 1-3. A schematic view of DNA replication. DNA is a double-stranded molecule, and the two strands of nucleotides are complementary. Because of their sizes and chemical properties, adenine is always opposite thymine, and guanine is always opposite cytosine. The double strand is not actually straight and flat as shown here, but twisted into a helix. When DNA replicates, the helix unwinds and the strands separate; nucleotides from the surrounding medium are then attached to their complements on the newly single strands, gradually building up two double strands identical to the original.

erally much shorter than DNA, are also polymers of nucleotides, but the sugar is ribose instead of deoxyribose, and the base uracil replaces the base thymine.

The precise sequence of bases along a stretch of DNA molecule is what constitutes genetic information. The chief function of this information is to specify the compostion of proteins. More than anything else, proteins make an organism what it is. Some are structural—collagen, for example, gives strength to skin and bone. The coordinated movements of other proteins underlie the contraction of muscle. Some proteins are carriers —hemoglobin, for example, carries oxygen from the lungs to the tissues of the body, and waste gas (carbon dioxide) back to the lungs. Perhaps the most fundamental to organisms in all the kingdoms of life is the role that proteins play as enzymes, or biological catalysts. Without enzymes, many essential biological reactions take place very slowly—some so slowly that they can hardly be said to proceed at all.

Like DNA and RNA, proteins are polymers, chains of small units. In proteins, the units are amino acids, small nitrogen-containing organic compounds. Only about twenty different ones, linked in chains from a few dozen to several hundred long, make up the proteins in all known organisms on Earth. A protein molecule consists of one or more amino acid chains twisted and folded into a shape that determines the protein's function. This shape is mainly determined by the amino acid sequence itself, although it can also be influenced by the presence of other molecules. A living cell synthesizes each chain by linking amino acids in an order precisely determined by one of its genes (see Figure 1-4). The code for translating the sequence of bases in DNA into a sequence of amino acids in a protein is nearly universal—a given base sequence would translate into the same amino acid sequence in almost all cases.

The precise copying of DNA from parent to offspring, although crucial to individual development, is not sufficient to ensure evolutionary change; there must also be a source of variety. If copying were always faithful, offspring cells would always be exact duplicates of their parent cells. If, then, environmental conditions altered so that the parent type was no longer

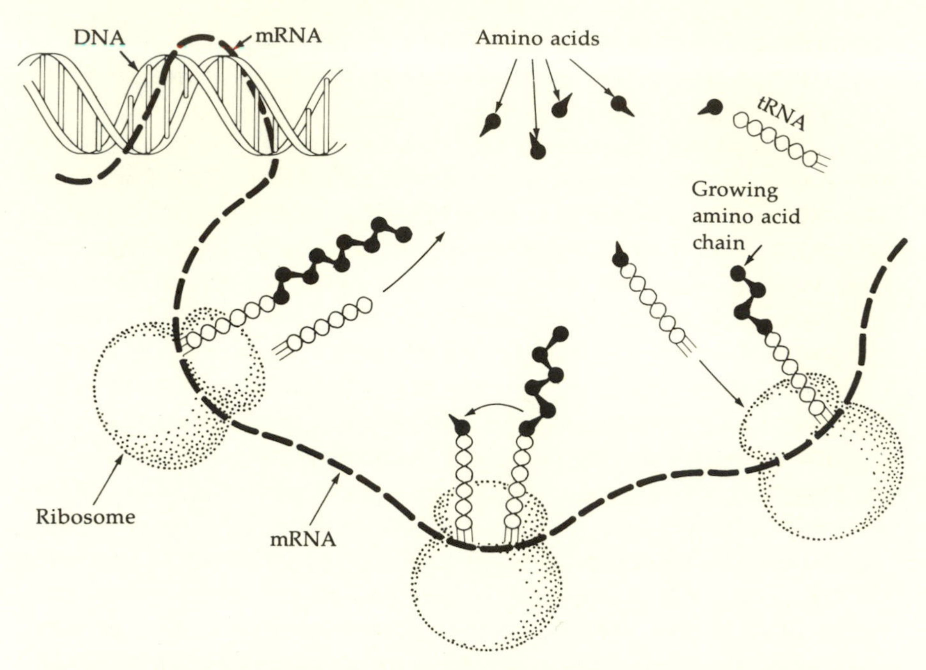

Figure 1-4. A schematic view of protein synthesis in a cell. First, a stretch of DNA produces a single complementary strand of messenger RNA (mRNA) by a process similar to DNA replication (see Figure 1-3). Amino acids are then linked into chains on ribosomes, somewhat top-shaped bodies made of rRNA (ribosomal RNA) and proteins. A ribosome attaches itself to the free end of an mRNA strand and begins to move along the strand. It "reads" the strand three nucleotides at a time—each such triplet codes for an amino acid molecule—and attaches the indicated amino acids to a growing chain of them. Amino acids are brought to the ribosome by tRNA (transfer RNA) molecules, of which there is a different kind for each of the twenty different amino acids that make up proteins.

suited to the job of making more of itself, the living, replicating system would die out. Among sexual organisms (those with two parents), variation comes mostly from a shuffling of the genetic material, for the offspring receives half its genes from one parent and half from the other. In Chapter 5, I shall describe the evolution of sexual strategies among the earliest eukaryotes. Among the asexual prokaryotes, however, change is usually the result of mutation, a change in the sequence of bases in a DNA molecule.

Mutations arise from a mistake in the copying process—substitution of one base for another or deletion of one or more bases—often as a result of damage to the DNA molecule by X-rays, ultraviolet radiation, or certain chemicals. Most mutations are harmful, resulting in the death or reduced reproductive capacity of the individual carrying the mutation. Occasionally, however, beneficial mutations arise, and these alterations will survive and be copied.

From the evolutionary point of view, it is not individuals, but populations of organisms that change. They adapt to changing conditions or they leave fewer and fewer copies of themselves and eventually become extinct. There is not enough room on the surface of the Earth, nor are there sufficient resources, for all the offspring produced by all organisms to survive and reproduce. Those organisms best adapted for survival and reproduction at the times and in the places they live leave more surviving offspring. Thus, certain genes—those in the survivors—pass on, while other genes are lost. These inexorable laws work to produce all forms of life, whether modern animal and plant populations or the most ancient populations of microbes.

To reproduce, as well as grow and do all the other processes required for life, organisms need energy, which heterotrophs derive from food. The chemical transformation of food into energy takes place within cells. In brief, a food molecule is broken down in a series of chemical reactions, and the energy released in the reactions is used to generate special "energy-rich" molecules that serve as the cell's immediate energy donors. The most important of these special molecules is adenosine triphosphate (ATP). Every living organism produces ATP, which is the immediate source of energy for virtually all cell functions.

A look at the structure of the ATP molecule will provide a clue to its role as a universal energy carrier. The molecule consists of three kinds of components: a nitrogen base (adenine), a sugar (ribose), and a sequence of three phosphate groups (see Figure 1-5). Under certain conditions, the other phosphate groups of the ATP molecule will split off. Because the bonds connecting the two outer phosphate groups to the molecule are "energy-rich" bonds, the splitting of ATP generates a large amount of usable energy. In the chemical reactions that occur in

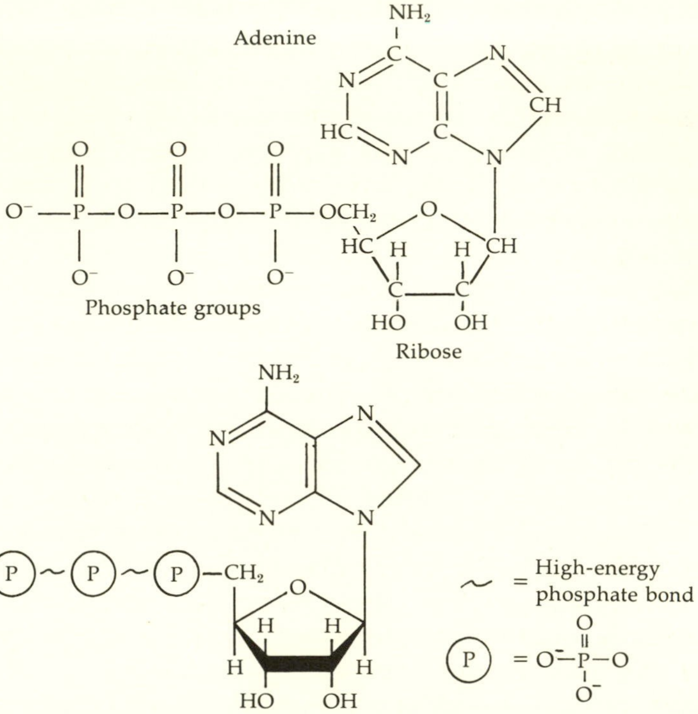

Figure 1-5. Two ways of representing the chemical structure of ATP.

cells, a phosphate group does not simply break off; rather, it is transferred with its energy to some other molecule. This energized molecule is then able to react with others in some way; it eventually does so and releases the phosphate group. The removal of one phosphate group from ATP leaves the less energetic molecule adenosine diphosphate (ADP); the transfer of two groups produces adenosine monophosphate (AMP). (The innermost phosphate group is anchored to the molecule by a firm bond that does not readily break.) Conversely, addition of one or two phosphate groups to AMP restores the energy-rich ADP or ATP.

Because cells require a continual supply of energy, they must continually replenish their store of ATP. Cells have evolved three major pathways for generating ATP: fermentation, respiration, and the light reactions of photosynthesis. These pathways and how they might have evolved are described in Chapters 2 and 3. Once the ATP-generating pathways evolved, they never died out. One of the most revealing and exciting insights of modern molecular biology is the extreme conservatism of metabolic systems. Not only have individual genes and the proteins made under their direction been conserved, but entire metabolic patterns have persisted for aeons. Organisms have added new acts to their repertoires only by rather superficial modifications. In particular, the fermentation, respiration, and photosynthetic pathways that evolved in prokaryotes several billion years ago are still the main ATP-generating pathways. The enzymes that catalyze these reactions are also ancient; current studies indicate that the enzymes have changed little since they appeared in early prokaryotes.

Reconstructing the Ancient World

What kind of world did the early prokaryotes inhabit? Were the conditions of the planet and its atmosphere those we know today? The Earth's surface, oceans, and atmosphere have been so profoundly altered by the activities of living forms on the planet that to answer these questions, one must turn to studies of our lifeless neighboring planets. The Earth condensed out of a cloud of dust and gases that formed it and the other planets of our solar system. Astrophysicists postulate that most of the major bodies of the solar system originated during the same period, about five billion years ago. Photographs taken from orbiting spacecraft show similarly-cratered surfaces on the Moon, Mercury, Venus, and Mars and its moons. The fact that the oldest rocks taken from the Moon's surface, as well as meteorites found on Earth, are all about 4.5 billion years old also supports the idea of a common origin of the major bodies of the solar system.

One can, therefore, consider Mars and Venus as sterile Earth-like places with similar planetary histories (the results of the Russian Venera 9 and 10 probes of Venus, as well as the United States' Viking mission to Mars in 1976, also suggest that this assumption is plausible) and make some good guesses about how life has modified the surface of our planet. One of the most conspicuous differences between the Earth and its neighbors is the large concentration of oxygen found in the Earth's atmosphere. The atmospheres of both Venus and Mars contain about 98 percent carbon dioxide and less than 1 percent oxygen (they also have about 2 percent nitrogen and some water vapor), whereas the Earth today has nearly 21 percent oxygen and only 0.03 percent carbon dioxide (and 79 percent nitrogen). When the Earth first formed, its atmosphere probably resembled the atmospheres of its neighboring planets at that time.

Biological considerations also support the theory that the young Earth's atmosphere contained no free oxygen. Life originated on the Earth through the formation and interaction of prebiotic compounds: nonbiologically produced amino acids, nucleotides, and sugars. Such chemical compounds simply do not accumulate in the presence of oxygen, which reacts with them and destroys them as soon as they form. The first cells on Earth, then, must have arisen in the absence of oxygen.

Primitive bacteria—those believed to be most directly descended from our earliest ancestor cells—are poisoned by oxygen. They have no chemical or other means of protection against the gas, and their cell material burns up, in effect, if exposed to it. Such primitive cells (called obligate anaerobes) live by fermentation, taking up organic compounds and generating ATP anaerobically. It is reasonable to assume that they evolved in the absence of oxygen.

In time, the supply of organic compounds became limited; the evolution of photosynthetic apparatus, which enabled cells to manufacture the organic compounds they needed from inorganic compounds using light as energy, occurred. The first photosynthesizers, however, were also anaerobic bacteria; their primitive form of photosynthesis did not generate oxygen.

How, then, did the Earth's atmosphere become oxygenic—a transition totally unpredictable from the laws of chemistry and physics? And when did the transition occur? To answer these questions, we have to look at the organisms that succeeded the earliest photosynthetic bacteria. These successors were the blue-green algae, a misnamed group of photosynthesizers that are not algae, nor are they always blue-green in color. Today, in recognition of the essential affinity of these microbes with other bacteria, biologists are beginning to use the term *blue-green bacteria,* or *cyanobacteria,* a practice I shall follow here. The cyanobacteria were probably the first organisms to give off oxygen as a waste product of their photosynthesis.

There is direct fossil evidence for the proliferation and diversification of cyanobacteria about 2.5 billion years ago, a date that fits nicely with evidence from the geological record, which shows two-billion-year-old rocks containing oxidized forms of minerals. The upsurge of oxygen in the Earth's atmosphere must have been due, then, to the worldwide proliferation of these bacteria. Never before or since have organisms on Earth so profoundly affected its atmosphere.

Because oxygen was toxic to early life, it became an increasingly serious pollutant. Like automobile waste products, this pollutant even threatened the producers themselves, the cyanobacteria. The resolution of the oxygen crisis was a turning point in the history of the cell: microbes evolved the capacity to use in respiration the oxygen that they produced. This solution not only protected them; it also provided them with additional energy, because respiration generates far more ATP than fermentation does. In time, as the concentration of atmospheric oxygen rose, cells of many nonphotosynthetic species evolved that required oxygen for their metabolic processes; these were the first obligate aerobes. They put the potentially poisonous oxygen to use in the elegant innovation of aerobic respiration. By this means, cells could generate enough ATP to grow larger and perform more sophisticated functions. About 600 million years ago, at the beginning of the Cambrian geological period, there was a virtual explosion of large forms of animal and photosynthetic life, their visible success the result of the miniaturized achievements of their microscopic ancestors.

For generations, the Cambrian rocks were thought to be the beginning of the fossil record. The time before the Cambrian was listed as a vast undifferentiated era, the "Precambrian," on geological time charts. Now enough is known about those times to recognize divisions in the Precambrian: the Hadean, the Archaean, and the Proterozoic Aeons (see Figure 1-6). The Hadean, whose name derives from Hades, the hot and chaotic underworld of Greek mythology, extended from 4.6 billion to 3.8 billion years ago. During this time, the Earth and its Moon took form as solid bodies. Meteorites and lunar rocks date from this period, but so much debris hit the Earth and there was such shuffling and melting of material on its surface that no terrestrial rocks remain from the Hadean. The Archaean Aeon, extending from about 3.8 billion to 2.6 billion years ago, saw the formation of the Earth's long-lived crustal features, the appearance of life on the planet, and development of the major metabolic strategies, including fermentation, photosynthesis, and the ability to convert atmospheric nitrogen to a form usable by cells. The beginning of the Proterozoic Aeon, about 2.6 billion years ago, is marked by a change in the character of the surface rocks. This aeon extended to the beginning of the Cambrian Period, about 600 million years ago. During the Proterozoic, eukaryotic cells developed two-parent sexual reproduction, giving rise to animal and plant ancestors. Eukaryotes of many forms evolved; by about a billion years

Figure 1-6. A broad chronology of Earth history. The major biological events of the Archean and Proterozoic aeons will be described in the following chapters.

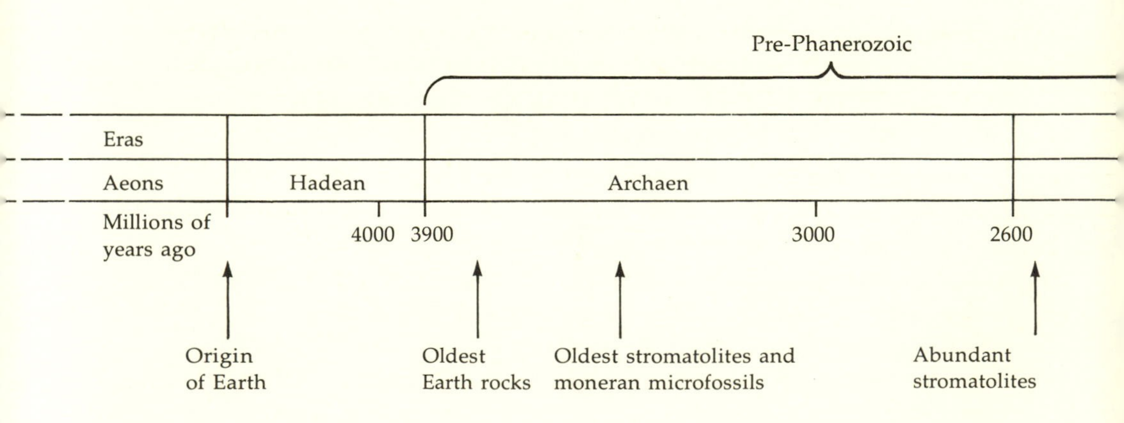

ago, large algae several centimeters in diameter had evolved; and near the very end of the aeon, about 700 million years ago, the first soft-bodied multicellular animals appeared. In keeping with this scale of time divisions, the time represented by the "classical" fossil record—trilobites, the first land plants and animals, the vast forests whose remains form our coal beds, the dinosaurs and woolly mammoths—has been named the Phanerozoic Aeon.

In this book, I shall depict the long quiet evolutionary build-up during the pre-Phanerozoic, especially the steps that led from single-celled anaerobic prokaryotes to the sophisticated eukaryotic cells whose existence was prerequisite to the origin and evolution of all large plants and animals. The backdrop for the narrative is the Earth's environment, which set the evolutionary processes in motion and then was transformed by them as the life forms in it changed. First, let us look at the world of our earliest ancestors and the strategies for survival that they developed billions of years ago.

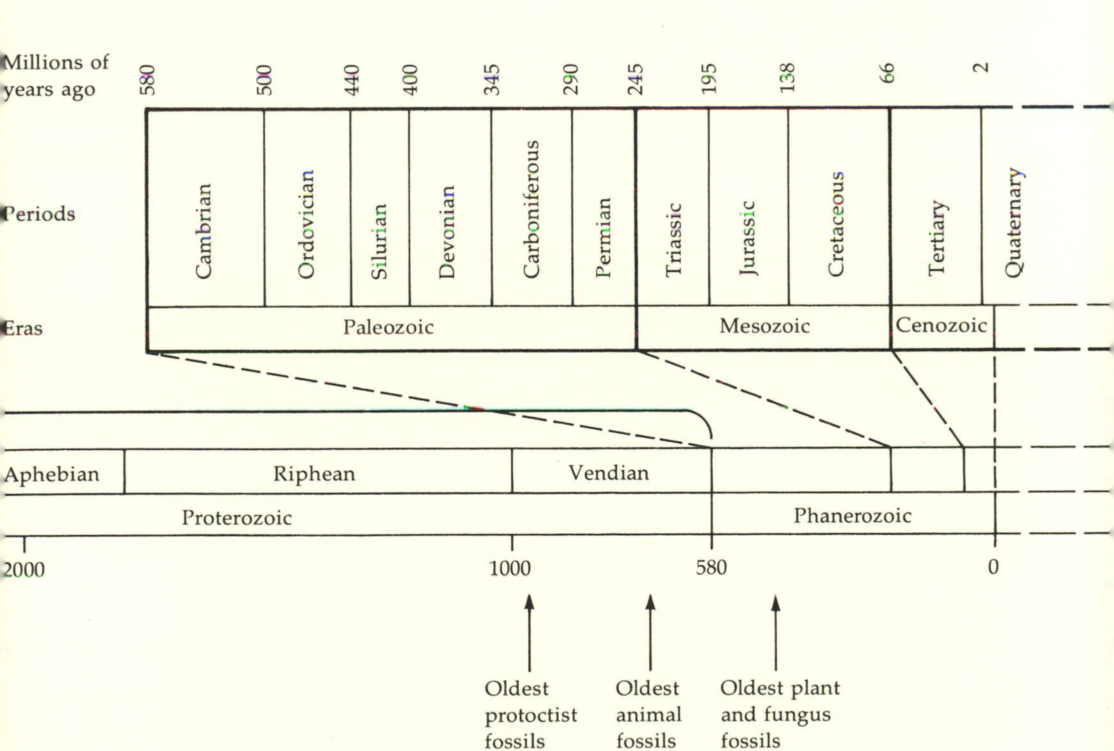

Suggested Reading

Broda, E. *The Evolution of the Bioenergetic Process.* Oxford: Pergamon Press, 1975.

Goldsmith, D., and T. Owen. *The Search for Life in the Universe.* Menlo Park, Calif.: Benjamin/Cummings, 1980.

Keeton, W. T. *Biological Science,* 3rd ed. New York: W. W. Norton, 1980.

Lovelock, J. E. *Gaia: A New Look at Life on Earth.* Oxford: Oxford University Press, 1979.

Margulis, L., and K. Schwartz. *Five Kingdoms: An Illustrated Guide to the Phyla of Life on Earth.* San Francisco: W. H. Freeman and Co., 1982.

Press, F., and R. Siever. *Earth,* 2nd ed. San Francisco: Freeman and Co., 1979.

CHAPTER 2

Life Without Oxygen

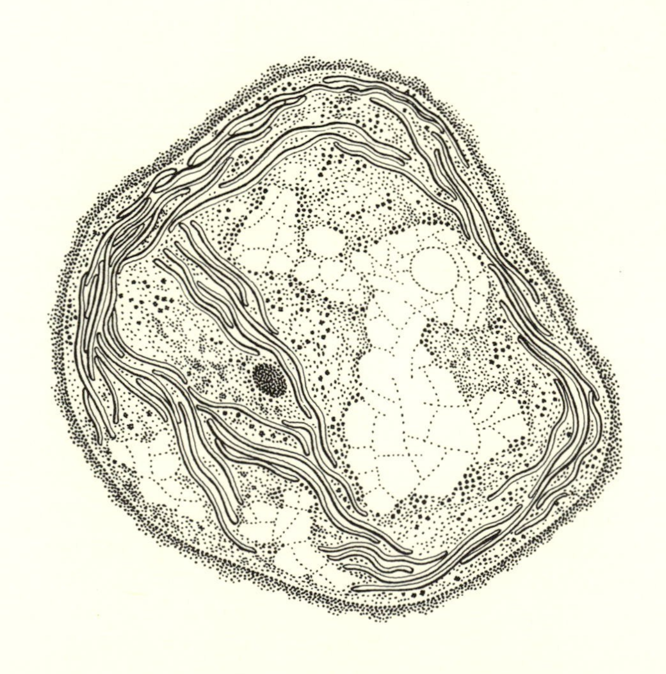

THE EARLY history of life on Earth is part of the early history of the planet itself, its surface, and its atmosphere. That history began about five billion years ago with the formation of the solar system out of a massive cloud of dust and gas, the solar nebula. The particles making up the solar nebula gravitated toward one another, and the cloud collapsed, spinning rapidly and flattening to form a disk. As gravitational collapse continued and inward pressure increased, the material near the center of the disk became denser and hotter. Eventually, the temperature became so great that the nuclei of colliding atoms (largely hydrogen, some helium) fused and thermonuclear reactions were ignited. The Sun was born as a star.

Not all the material in the solar nebula drifted toward the center. While the Sun was still forming, dust particles and gases at some distance from the center began to accrete and eventually formed planets bound in orbits around the Sun. One of these planets was the Earth. The history of the Earth's first billion years—the Hadean Aeon—can only be inferred from the models of astronomers and from studies of the Moon and meteorites. The planet suffered so much bombardment by other objects in the solar system and so much movement of material on its surface that no intact sample of rock survived the cataclysmic, primordial events. It seems that the ancient Greeks were right: the Earth was born in chaos.

As the newly formed Earth grew, it began to heat up. Enormous amounts of heat must have been released as the inner parts of the growing mass were compressed by the accumulation of the outer layers. Meteorites and other planetary debris released additional heat as they hit with sufficient impact to form huge craters. But probably the most significant source of heat on the new planet came then, as it still does, from the decay of radioactive elements (uranium, thorium, and potassium) in its interior.

Heating had a profound effect on the ultimate structure of the Earth: it led to the differentiation of its rocky substance into concentric layers. As the interior became hotter, the rocks melted. The heavier elements, iron and nickel, sank to the inside, forming a liquid core. The lighter ones, such as aluminum, silicon, oxygen, and magnesium, floated to the top and reacted with each other to form silicate minerals and rocks that made up the outer layers. (Silicates are a group of minerals containing silicon, oxygen, and various metals.) The differentiation was not complete. Some of the light elements that rose to the top had formed chemical bonds with heavier ones such as iron and the radioactive elements uranium and thorium. Thus, heavy elements are still to be found near the Earth's surface.

Unreactive gases, such as nitrogen, argon, and neon, did not enter into chemical combination with other elements, although they were retained by the gravity of the planet. They therefore became part of the outer surface, the primary atmosphere. This atmosphere did not last long, however. The onset of thermonuclear reactions in the Sun led to a sudden and tremendous increase in outward pressure, which caused a mass of material to be blown into space. The violent stream of material from the Sun was so energetic that it blasted away the primary atmospheres of the nearest planets, including that of the Earth.

The melting and differentiation of the Earth initiated vulcanism. Lava spewed forth and spread to form a thin crust. This primordial crust was too thin to be stable; it remelted and solidified repeatedly. As the planet began to cool, however, the crust solidified more permanently into a rocky layer composed largely of aluminosilicates. Meanwhile, emissions from the planet's molten interior gradually formed an atmosphere to replace the one that had been blown away when the Sun "turned on." Water vapor, nitrogen, argon, neon, carbon dioxide, and hydrogen now surrounded the Earth with a new, secondary atmosphere. On the hypothesis that the early history of Venus,

Earth, and Mars was similar, the recent exploration of these neighboring planets suggests that the Earth's secondary atmosphere, like their present ones, contained less than one percent oxygen, but larger quantities of nitrogen and carbon dioxide (perhaps as much as 10 percent of the atmosphere was CO_2). It is likely that the atmosphere also contained small quantities of hydrogen and hydrogen-rich gases—ammonia (NH_3) and methane (CH_4)—but it could not have retained large amounts of these gases. Hydrogen, the lightest gas, would have escaped into space; so would some of the methane. Methane would also have been oxidized, as apparently it has been on Mars and Venus, and ammonia would have dissolved in water or been destroyed in the atmosphere by solar ultraviolet light.

While the Earth was forming, water had been "locked up," chemically bound in minerals as water of hydration. The great heat produced in the differentiation of the Earth broke these chemical bonds, releasing the water molecules. Steam exploded from volcanoes, condensed, and fell back as rain. As the Earth cooled, water condensed in increasing quantity, eventually forming the hydrosphere—ground water, oceans, lakes, rivers, streams, ponds, hot springs, and geysers.

By the start of the Archaean Aeon, some billion years after the formation of the planet, the Earth probably had continents, shallow basins of water, and an over-all cratered and pockmarked surface; it probably resembled a wet Mars. As inhospitable as such an environment might seem now, it provided the raw materials for life's beginnings.

In our solar system, the Earth is uniquely suited for the formation of life as we know it: carbon-based chemical systems in salt water. The Earth is the right distance from the Sun. Life cannot exist outside the range of temperature in which water is liquid; if water is present only as solid ice or as steam, living systems will not form or survive. On planets too far from the Sun, such as Mars, temperatures are too low to permit the formation of open bodies of liquid water. Conversely, Venus is too close to the Sun: what tiny amounts of water it has are in the form of steam.

The Earth is also the right size. If it were much smaller, it would not have enough gravity to hold an atmosphere; and planets without atmospheres are unable to cycle elements through a hydrosphere, or fluid medium. Conversely, if the Earth were much larger, its atmosphere would be so dense that solar radiation—the ultimate source of the energy required for life's chemical reactions—would be unable to penetrate to the surface of the planet.

The Beginning of Life

How did life begin on Earth? How could even the simplest cell evolve from nonliving chemical precursors? These are formidable research questions. Eventually, it may be possible to specify the steps that can lead to the formation of DNA, RNA, and proteins packaged in a lipid/protein membrane and organized in such a way as to ensure their replication. It may actually be possible, from biochemical considerations alone, to reconstruct conditions at the surface of the Archaean Earth—its temperature, gas composition and pressure, salinity, humidity, and surface minerals—as well as tidal, daily, and seasonal changes in these conditions.

Although the picture still lacks many details, there is broad agreement on certain general assumptions. One is that life arose by the self-assembly of small organic molecules into larger, more complex molecules. According to one hypothesis, the assembly took place on the surface of clays or other crystals. Such ordered surfaces could have attracted small molecules and held them in arrays that promoted the formation of small polymers—short chains of amino acids or nucleotides. The existence of these short chains would have made it possible for larger structures—primitive proteins and nucleic acids—to form spontaneously. The building blocks themselves, the small organic compounds, would have been formed by the action of lightning, solar ultraviolet radiation, and other forms of energy on the gases of the secondary atmosphere.

The development of an experimental approach to the study of the origin of life must be credited to a short monograph published in 1924 by the Russian biochemist Alexander I. Oparin and to an article published in 1929 by the British physiologist J. B. S. Haldane.* Oparin and Haldane pointed out that an atmo-

*See Suggested Reading at the end of this chapter.

sphere lacking free oxygen was a requirement for the evolution of life from nonliving organic matter; oxygen would have reacted instantly with any prebiotic (nonbiologically produced) organic compounds or would have prevented them from forming at all. It was not until the 1950s, however, that Stanley L. Miller and Harold C. Urey, then at the University of Chicago, demonstrated that organic compounds could be synthesized from the gases thought to have been present on the young Earth. In their experiments, Miller and Urey discharged an electric spark through a mixture of such gases in a glass vessel. Since then, many similar experiments have produced organic compounds—including amino acids and nucleic-acid bases—from various mixtures of simple gases. Several forms of energy are effective—electric spark, silent electric discharge, ultraviolet radiation, and heat. There is no doubt that the young Earth had a supply of organic compounds suitable for the assembly and sustenance of the first living organisms.

When did life arise on Earth? The recognition of the meaning of fossils in the nineteenth century removed discussion of life's origins from the realms of mythology and theology. This enabled the discoverers of animal fossils to work out a chronology that placed life's origins at about 600 million years ago, just before the Cambrian Period. This view prevailed until the 1950s. Most paleontologists then believed that the earliest life had been soft-bodied single-celled animals, immediate ancestors of the trilobites, worms, and other marine bottom-dwelling animals predominant in the early Cambrian (see Figure 2-1). Textbooks published before about 1960, as well as exhibits still found in museums around the world, explain that at the beginning of the Cambrian the fossil record shows an explosion of well-developed animal remains on several continents, with few or no precursors.

It is easy to understand why the story of life before the Cambrian did not emerge for so long. Until well into the twentieth century, the best-studied fossil deposits were those of England and Wales, where the Cambrian rocks overlie much older rocks. The boundary between these rock layers is a sharp discontinuity—it represents a time interval in which deposition of sediment stopped and erosion removed some of the sediment. In some places, millions or even hundreds of millions of years elapsed between the deposition of lower and upper rock layers.

Figure 2-1. The beginning of the Phanerozoic Aeon: a marine landscape of the Cambrian Period, some 600 million years ago. [Drawing by Laszlo Meszoly.]

Because there are no obvious well-preserved fossils in the lower, pre-Cambrian rocks, the British Cambrian/pre-Cambrian discontinuity had given the impression that the beginning of the Cambrian itself represented an enormous discontinuity, perhaps preceded directly by the origin of life.

In the 1950s, Martin Glaessner, of the University of Adelaide, found a rock sequence in Ediacara, South Australia, that showed no discontinuity between Cambrian and late pre-Cambrian sediments. And below the Cambrian rocks were excellent impressions left in sand or mud by soft-bodied marine animals of many recognizable forms (see Figure 2-2). Since Glaessner's discovery, Ediacaran fauna (as similar fossils are now called) have been found in several other places, including England, Greenland, and Siberia. It is now the consensus that the problem of the suddenness of the appearance of life was due partly to the nature of the evidence (see Figures 2-3 and 2-4). The soft parts of organisms are fossilized much more rarely than the

Figure 2-2. The end of the Proterozoic Aeon: an Ediacaran sea bottom with soft-bodied animals reconstructed from the impressions that they left in sand and mud some 700 million years ago. [Drawing by Laszlo Meszoly.]

hard parts are, and hard parts, primarily the calcium phosphate and calcium carbonate shells of invertebrate animals, first appeared only just before the Cambrian. They then became widespread after some tens of millions of years—rather suddenly, to be sure, from our perspective of some half a billion years.

Evidence of microbial predecessors to the Ediacaran fauna was first revealed by the paleontologists Elso S. Barghoorn, of Harvard University, and Stanley A. Tyler, of the University of Wisconsin. Acting on the hunch that microbial life ought to have preceded the appearance of large animals and plants, in the mid-1950s Barghoorn and Tyler undertook microscopic studies of certain ancient sedimentary rocks about two billion years old, part of the Gunflint Iron Formation of Ontario and northern Minnesota. The researchers were right. The Gunflint rocks abounded in microscopic forms, and some of them looked enough like modern bacterial cells to be fossils of microbial life

Figure 2-3. The boundary between pre-Phanerozoic and Phanerozoic rocks in the Monument Creek area of the Grand Canyon in Arizona. The discontinuity here is profound. Because no obvious signs of life are found below the arrow and fossils are abundant above (in the Cambrian strata, which begin with the Tapeats sandstone), it was thought that this boundary represented some major change in the history of Earth and of life. [Photograph by Bradford Washburn/Boston Museum of Science.]

Figure 2-4. The lower boundary of the Cambrian rocks on the Aldan River, in Siberia. The arrow points to the lowermost rock layer, dolomite, that contains fossils of skeletalized Cambrian fauna. The layers above it contain fossils of more than sixty species of skeletalized animals. The layers below, also dolomite, lack such remains but contain stromatolites (sedimentary rocks laid down by communities of bacteria) and traces of soft-bodied Ediacaran fauna. The lack of discontinuity between these rock layers supports the notion that the lower Cambrian boundary represents a change in the preservability of fossils, an evolutionary event, rather than a geological, atmospheric, or other environmental event. [Courtesy of Max and Françoise Debrenne, Musée d'histoire naturelle, Paris.]

(see Figure 2-5). Since the early 1960s, there have been many discoveries comparable to the Gunflint find. The oldest, from sedimentary rocks of the Warrawoona Series at North Pole, Western Australia, were discovered in the late 1970s by Stanley M. Awramik, of the University of California at Santa Barbara. They were described by him and the Precambrian Paleobiology Research Group, under the direction of J. Wm. Schopf, of the University of California at Los Angeles. The fossils are about 3.5 billion years old, but they are so complex in form that life must already have been well along when they were laid down (see Figure 2-6).*

The Invention of Fermentation

The first cells on Earth probably resembled the minimal organisms known today: tiny bacterial cells that are little more than membranous bags of water containing genes (DNA), soluble enzyme proteins, and ribosomes. Tiny bacteria of this sort have limited abilities to manufacture their own cell components. They lack sufficient genetic information to make all of the amino acids, nucleotides, vitamins, and sometimes even the enzymes and the larger proteins they require. They must receive their particular chemical needs as food. Nowadays, in fact, many are parasites, richly supplied with their needs by their animal or plant hosts.

How did the first living cells find the food they needed? Without the barrier of the ozone layer, the radiation reaching the Earth's surface from the Sun must have been intense. It was probably at least as intense as the radiation at the surface of Mars, whose atmosphere is transparent to ultraviolet light. In such an environment, the production of organic compounds would have flourished; food would have been manna from heaven, produced by sunlight. The first cells, then, probably lived as heterotrophs, able to feed on organic molecules similar to those of which they were composed: amino acids, sugars, and small organic acids.

Even the simplest cells, however, need both food and energy

*Because the age measurements cannot be made on precisely the same samples from which the fossils are recovered, these microfossils actually may be embedded in far younger (70 millon years old) rocks. The remarkable preservation and the immense importance of this discovery require us to use the utmost caution in interpreting it.

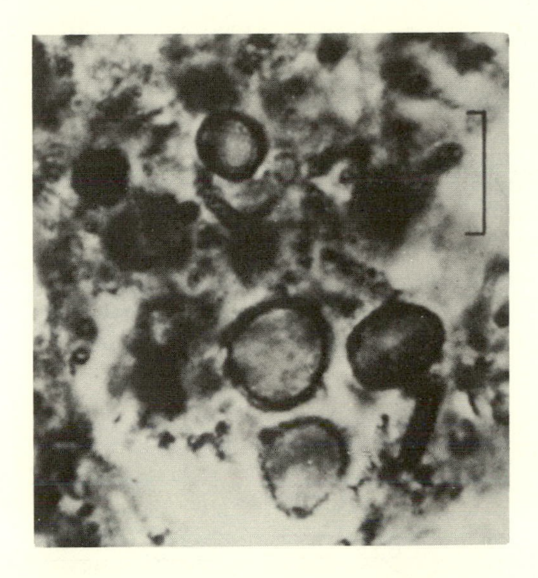

Figure 2-5. Microfossils in rocks of the Gunflint Iron formation, about two billion years old. [Courtesy of Stanley M. Awramik, University of California at Santa Barbara.]

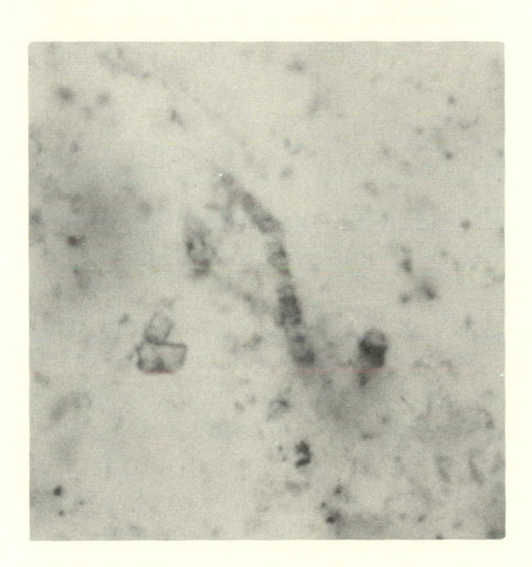

Figure 2-6. A complex microfossil, a bent filament having cross walls, from a rock found at North Pole, Western Australia. The filament is about five micrometers (five millionths of a meter) wide, the same width as filaments of modern blue-green algae, or cyanobacteria. The rock may be about 3.5 billion years old. (See footnote, page 32.) [Courtesy of Stanley M. Awramik, University of California at Santa Barbara.]

to grow and reproduce. They require energy in the form of ATP to produce DNA, RNA, and protein from organic precursors supplied by the environment. In experiments designed to simulate conditions of the early Archaean, researchers have readily obtained molecules of ATP from a variety of simple gas mixtures. This suggests that the first cells could have obtained energy in the most direct way possible: they simply ate ATP. Indeed, they probably ate a wide range of high-energy compounds, including two that are related to ATP—guanosine triphosphate and uridine triphosphate. The situation could not last. As increasing numbers of cells came to populate the Earth, the supply of such high-energy compounds must have become depleted. Cells then developed internal mechanisms for producing such compounds themselves. Probably one of the first of these mechanisms—to this day it is the most widely used—was fermentation.

Fermentation, which takes place in the absence of oxygen, is the breakdown of organic compounds (typically, sugars) into smaller organic compounds containing less energy. The released energy is used to generate ATP from ADP and phosphate (Ⓟ):

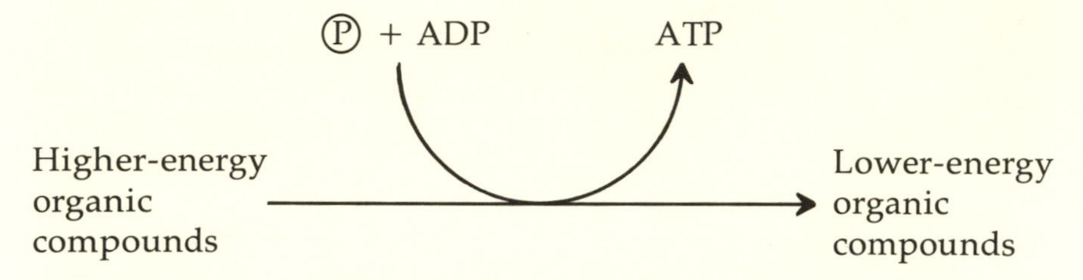

The breakdown takes place by a series of chemical reactions, and living organisms have developed numerous different fermentation pathways. However, many fermenting organisms use a common sequence called the Embden-Meyerhof pathway, after two of its discoverers (see Figure 2-7). The net effect of this pathway is glycolysis: glucose, a sugar molecule containing six carbon atoms, is broken down into pyruvic acid, whose molecules contain only three carbon atoms; two molecules of ATP are produced for every fermented molecule of glucose. Variations in fermentation by different organisms are chiefly in the pathways leading from various starting compounds to glucose, and in the fate of the pyruvic acid. In animal cells and some bacteria, pyruvic acid is converted to lactic acid when the absence of oxygen prevents respiration. In plant cells, yeasts, and some other bacteria, pyruvic acid is converted to carbon dioxide and ethanol under oxygen-poor conditions. This pathway is the basis of the wine, beer, and liquor industries. Many fermenting bacteria convert pyruvic acid into other end products, chiefly organic acids.

Fermenting bacteria abound in today's soils and waters; they are abundant too in sewage treatment plants, where they break down wastes, and in vinegar and wine vats, where they produce acetic acid and ethanol. So far as they have remained unchanged

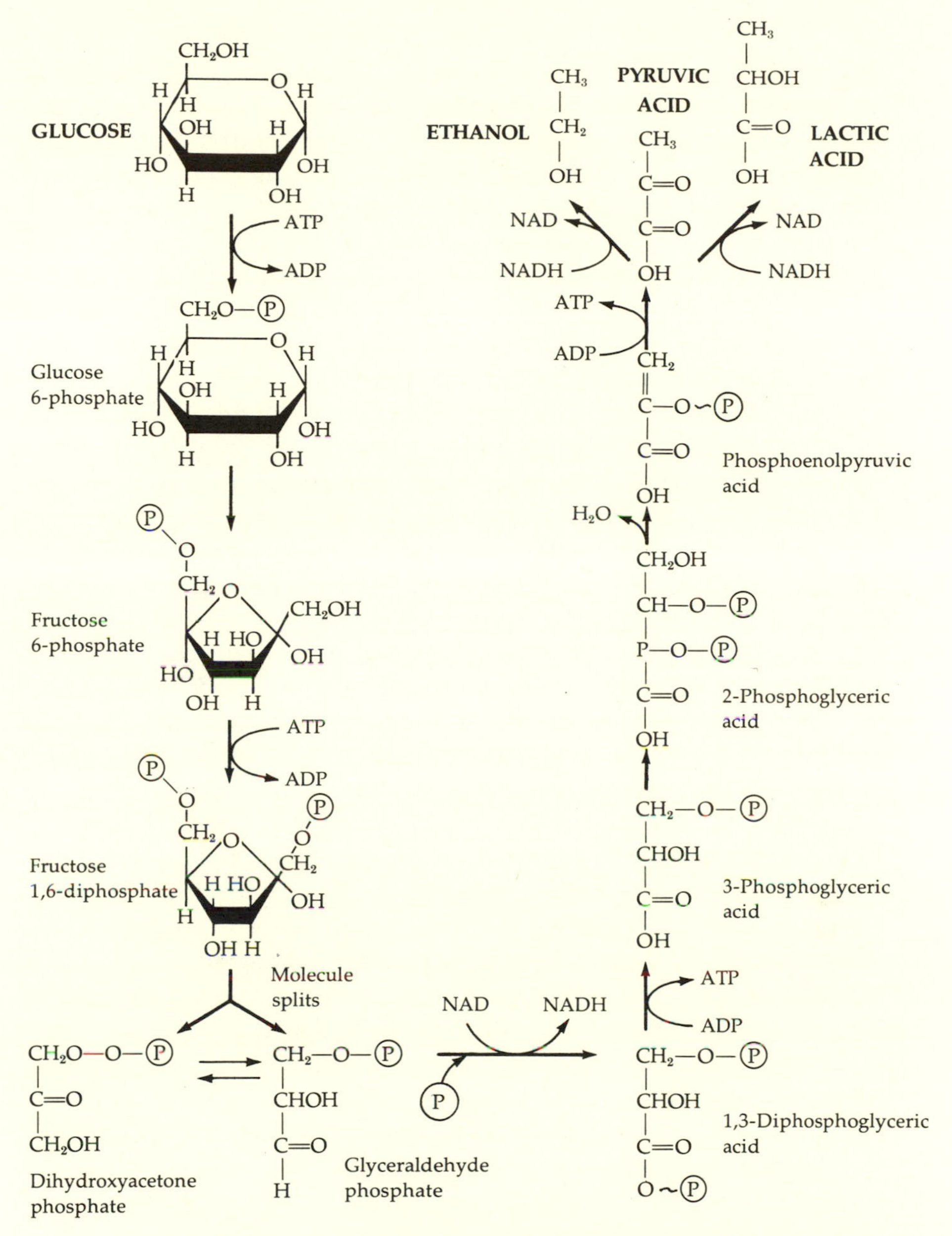

Figure 2-7. Glycolysis, the breakdown of glucose (a sugar) to pyruvic acid, is the heart of fermentation in many organisms. Each reaction is catalyzed by a different enzyme. The further transformations of pyruvic acid—only two possible ones are shown here, with several reaction steps omitted—vary widely among different organisms. Notice that the substance NAD (nicotinamide adenine dinucleotide) is recycled, being reduced to NADH and back again.

since their ancestors appeared in the Archaean, they are living models of the Archaean life style. Two major groups of fermenters, lactic-acid bacteria and clostridia, are thought most to resemble early cells. The lactic-acid bacteria (*Lactobacillus, Streptococcus,* and other genera) typically produce lactic acid as an end product of fermentation; they are the bacteria that sour milk and sauerkraut and ripen some cheeses. The clostridia (all members of the large genus *Clostridium*) live in soil, dust, water, and the gastrointestinal tract of animals, including human beings, in whom they can cause serious diseases. Many lactic-acid bacteria can tolerate exposure to oxygen, but they make no metabolic use of it; clostridia cannot grow in the presence of oxygen. Different species of clostridia and lactic-acid bacteria have evolved many different ways of making a living by fermentation. Typically, they start with some sugar—sucrose, glucose, lactose, mannose, or arabinose—or with a compound such as cellulose or starch, composed of linked sugar molecules. Other starting compounds include amino acids, alcohols (such as ethanol, ribitol, and mannitol), and acids (such as linoleic acid, a component of fats).

Fermentation is an inefficient form of metabolism. The end products, which pass as excretions into the environs, still contain energy. As time passed, microbes evolved that could make use of the different energetic waste products of the early fermenters. The fermentation end products of one species became the starting compounds—or food—for another. Such fermentation food chains still exist in swamp and lake muds. For example, acetic acid bacteria—sugar fermenters—generate acetic acid as a waste product of their fermentation. Other bacteria convert the acetic acid to lactic acid or to carbon dioxide and water.

In addition to numerous fermentation pathways, many other metabolic strategies have been evolved by the clostridia. Perhaps the most significant of these is nitrogen fixation, the incorporation of nitrogen from the atmosphere into the organic nitrogen compounds of cells. In this process, inert atmospheric nitrogen is first converted into ammonia

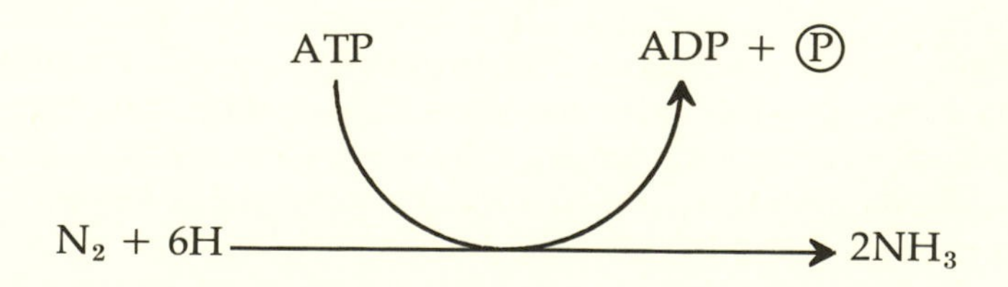

and then into nitrate (NO_3^-). These nitrogen compounds eventually come to constitute the amino acids of proteins and the nitrogen-containing bases of DNA and RNA. In fermentation, small amounts of these compounds are degraded to nitrogen gas. If the reverse conversion were not made, eventually most nitrogen would be in the atmosphere and life on Earth would die of nitrogen deficiency.

A molecule of nitrogen gas consists of two tightly interlocked atoms (N_2); to break the bond connecting these atoms requires an enormous input of energy. The industrial process for fixing atmospheric nitrogen (chiefly for fertilizer) takes place at 500°C and at least 300 times normal atmospheric pressure; biological fixation requires the expenditure of from 6 to 18 molecules of ATP for each molecule of nitrogen that is fixed. Not all organisms can perform this feat. In fact, no eukaryote is capable of it—all plants and animals depend for their existence on the fixing ability of certain prokaryotes. However, the clostridia are not the only prokaryotes that can fix nitrogen. Perhaps the best studied, because of their importance to agriculture, are bacteria of the genus *Rhizobium*, which benignly infect the roots of legumes (peas and beans). Cyanobacteria (blue-green algae) also fix nitrogen, as well as perform photosynthesis.

In all these organisms, nitrogen fixation is catalyzed by similar complexes of enzymes—the complex is called nitrogenase. It is supplied with hydrogen atoms by carrier molecules, iron- and sulfer-containing proteins called ferredoxins (see Figure 2-8). In clostridia, the hydrogen atoms are cleaved off

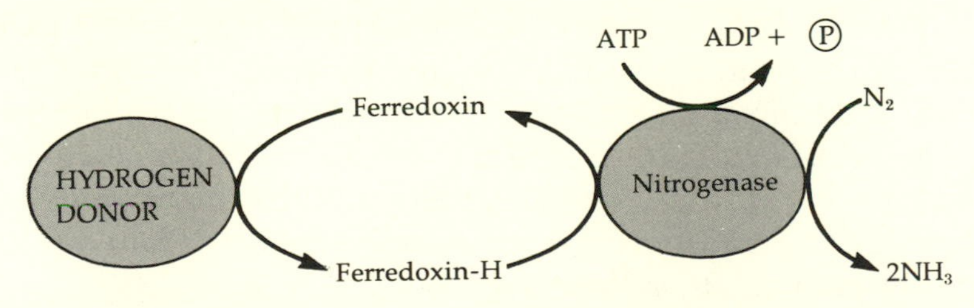

Figure 2-8. A schematic diagram of biological nitrogen fixation. Free ammonia (NH_3) is not actually produced. However, the splitting of the nitrogen molecule and the reaction of the nitrogen atoms with hydrogen are essential steps toward the production of nitrogen-containing organic compounds.

from organic compounds during fermentation, or are obtained directly from hydrogen gas in the sediment. In cyanobacteria, they are available from water molecules split during photosynthesis. As it happens, ferredoxins also play an important role in photosynthesis, but that was a later development. It was primitive fermenters, the early clostridia, that first evolved ferredoxins and nitrogenase. Today the clostridia, along with later evolved groups that possess this innovation of the Archaean, continue to supply the entire biosphere with organic nitrogen compounds.

Electron-transport Chains and Porphyrin Rings

Once cell reproduction had gotten well under way, the supply of chemical requisites for sustaining life set limits on the amount of living matter that could be made. Cells would have depleted food as fast as it could be generated by the Sun. If new, autotrophic food-producing strategies had not evolved, life might have ended at that point. At best, it would have remained sparse, in balance with the steady nonbiological production of organic food.

The first of the new metabolic pathways, one that can be viewed as a step on the path to autotrophy, evolved in a group of bacteria such as *Desulfovibrio*. Nowadays, desulfovibrios and similar bacteria (*Desulfotomaculum* and *Desulfuromonas*) live in anaerobic sediments containing organic matter and sulfate (SO_4^{2-})—in the mud at the bottom of ponds and streams, in bogs, and along the seashore. They ferment organic compounds such as lactic and pyruvic acids to acetic acid, and they convert sulfate into sulfide (S^{2-}). The sulfide may be released as hydrogen sulfide (H_2S) or dimethyl sulfide (H_3CSCH_3), smelly gases familiar from rotten eggs and decaying seaweed. Desulfovibrios and their kind can also ferment in the absence of sulfate, but they derive more ATP from their food if they can simultaneously convert sulfate into sulfide. This was their valuable innovation—to produce ATP by transferring high-energy electrons from the fermentation pathway to sulfate (see Figure 2-9).

Almost every chemical reaction entails a transfer of electrons. For example, when hydrogen and oxygen atoms combine to form water

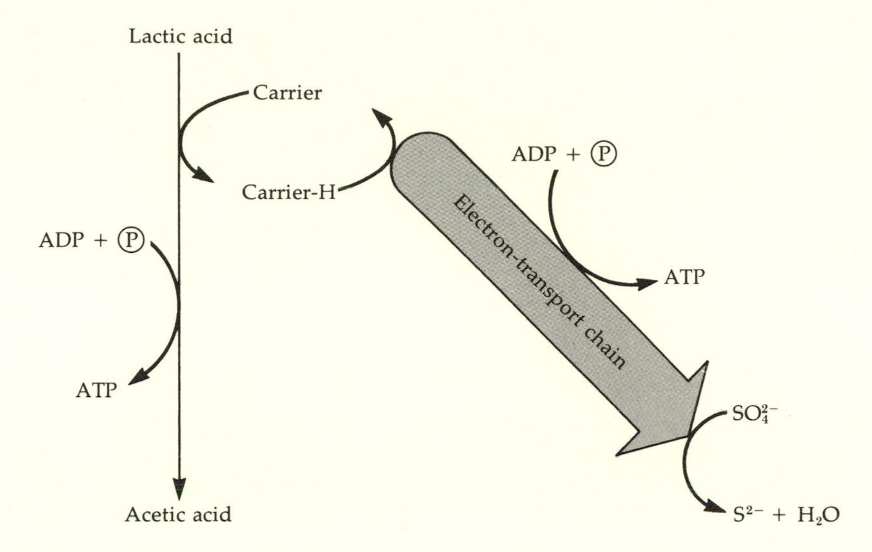

Figure 2-9. Some anaerobic bacteria not only ferment lactic to acetic acid (detailed steps omitted) but also produce ATP by reducing sulfate (SO_4^{2-}) to sulfide (S^{2-}). Because metal sulfides are almost insoluble in water, dissolved metals precipitate from solution in the presence of sulfide. Thus, such bacteria may have helped to form many of the world's sedimentary metal ore deposits.

$$2H + O \rightarrow H_2O$$

the single electron that each hydrogen atom brings to the association spends most of its time near the oxygen atom. It is conventional to say that the hydrogen atoms have been oxidized and the oxygen atom has been reduced. In general, reduction is the addition of electrons to an atom or molecule, and oxidation is their removal—oxygen itself need not be involved, and the transfer of electrons need not be complete. In fact, in most water molecules it is not—the electrons still spend some time near the hydrogen atoms. Occasionally, however, a water molecule completes the transfer; it dissociates, or splits up

$$H_2O \rightarrow H^+ + HO^-$$

into an electronless and thus positively-charged hydrogen atom (a proton, or hydrogen ion) and a negatively-charged hydroxyl ion. Any sample of water contains these ions. In the watery interior of a cell, electrons in transit (being negatively charged) are dogged by the ever-present hydrogen ions. Thus, in biological systems, reduction is equivalent to the addition of hydrogen atoms and oxidation to their removal. For example, in nitrogen-fixing bacteria, nitrogen (N_2) becomes reduced to ammonia (NH_3) and the source of hydrogen becomes oxidized.

Likewise, desulfovibrios reduce sulfate to sulfide and water, and they oxidize certain molecules in the fermentation pathway. High-energy electrons (or hydrogen atoms) do not pass directly from the oxidized organic molecule to sulfate, but move gradually along an electron-transport chain—a sort of bucket brigade of molecules that are capable of reversible oxidation and reduction. As a molecule in the chain receives an electron, it is reduced; the molecule is oxidized, returning to its former state, as it releases the electron to the next member of the chain. Each transfer of electrons is catalyzed by a specific enzyme, and the energy released by some of the transfers is used to make ATP. In desulfovibrios, the last member of the chain passes the electron to sulfate.

The invention of electron-transport chains has had incalculable consequences. They are an indispensable part of the photosynthetic machinery of bacteria, algae, and plants, and also of respiration, the ATP-producing pathway in all eukaryotes. In all organisms, electron-transport chains are remarkably similar in the molecules that make them up—both the enzymes and the electron carriers themselves. Some of them are small organic compounds containing no more than a few dozen carbon atoms. Others are proteins, such as ferredoxins and the cytochromes.

The electron-carrying capacity of a cytochrome is due to a porphyrin ring, a rather small ring-shaped compound attached to the protein chain of the cytochrome (see Figure 2-10). In cytochrome, the center of the porphyrin ring is occupied by an iron atom. This porphyrin-metal complex is called a heme group, present also in blood hemoglobin. Because of the alternation of single and double bonds around the heme group, electrons can travel rather freely within the molecule. In particular, they can

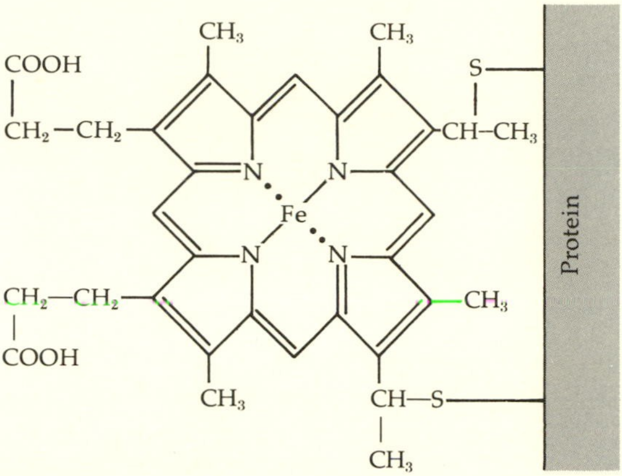

Figure 2-10. The heme group of cytochrome. The iron atom is ionized either as Fe^{3+} (ferric iron) or Fe^{2+} (ferrous iron). It changes back and forth by accepting and releasing electrons.

move between the iron atom and the edge of the ring. In cytochrome, the protein chain enfolds almost the entire heme group, which thus has only one exposed edge (see Figure 2-11). When a carrier molecule presents an electron to this edge, the electron moves rapidly to the center of the ring and reduces the iron atom. Likewise, a reduced iron atom can easily release an electron through the exposed edge of the heme group.

Porphyrin–protein complexes such as cytochromes must have appeared very early in the history of bacterial cells, possibly at the origin of the desulfovibrios themselves, but after the evolution of fermenting bacteria. Clostridia and other cells capable only of fermentation lack porphyrins.

The porphyrin ring was a prerequisite for the evolutionary development that freed cells totally from their dependence on preformed organic foods—photosynthesis. All photosynthetic organisms contain some form of chlorophyll, a green pigment* molecule composed of a magnesium-containing porphyrin ring

*Pigments are compounds that strongly absorb certain wavelengths of white light and, therefore, reflect the remainder of the visible spectrum. Because chlorophyll absorbs blue and red light, it reflects green.

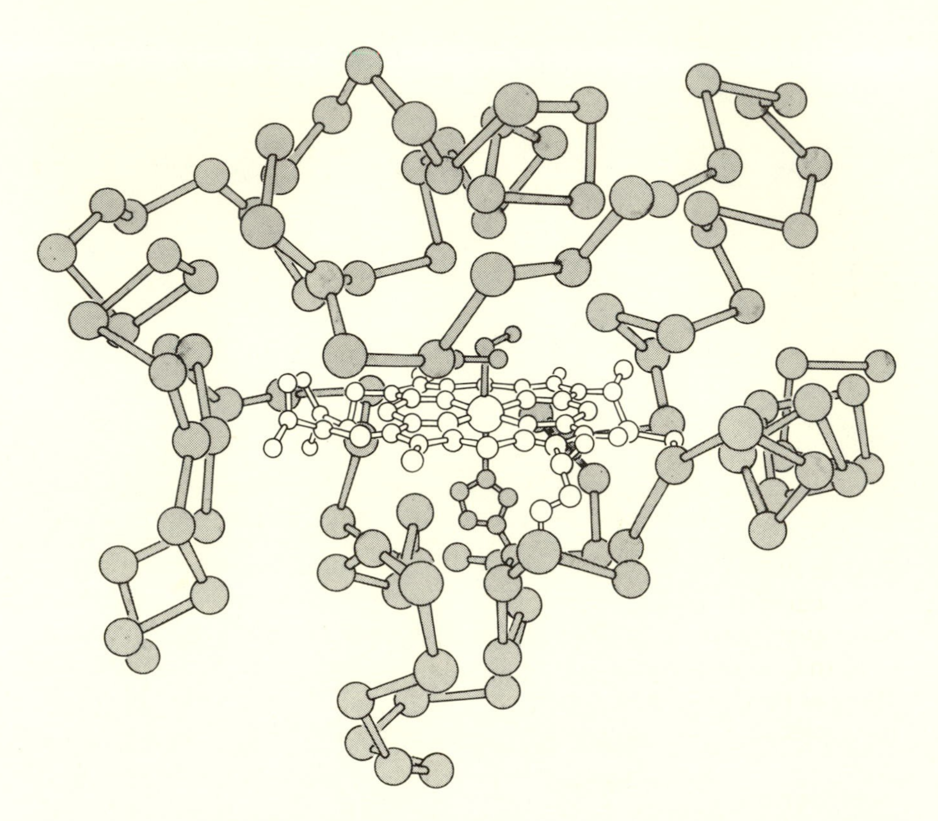

Figure 2-11. A model of the cytochrome *c* molecule found in tuna fish. The white spheres represent the atoms of the heme group. Each large dark sphere represents a whole amino acid of the cytochrome protein. The small dark spheres represent individual atoms of the amino acids on either side of the iron atom (large white sphere) in the center of the heme group. Although they differ in detail, the cytochrome *c* molecules in all organisms so far studied—whether bacteria, algae, or tuna fish—have a similar structure. [Drawing by Laszlo Meszoly.]

with a long tail (see Figure 2-12). The appearance of photosynthesis, therefore, was made possible by the prior evolution of magnesium porphyrins, as well as of cytochromes, which cells also need to carry out photosynthesis. The first photosynthetic organisms, then, evolved from bacteria that already contained the enzymes for synthesizing porphyrin–protein complexes, bacteria probably like the sulfate reducers.

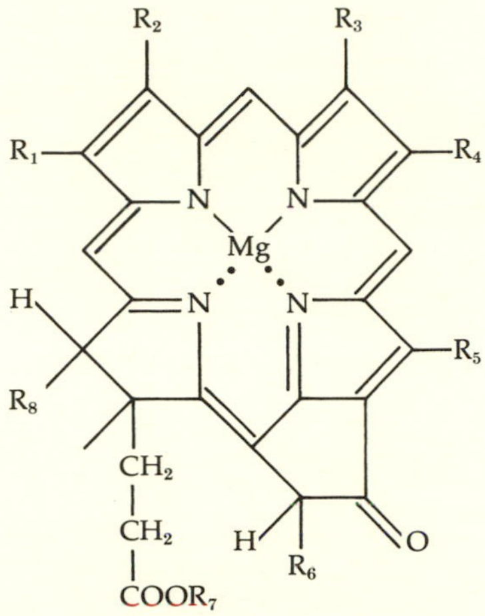

Figure 2-12. Chlorophyll is a porphyrin ring having a magnesium atom (Mg) in its center. Chlorophylls from different organisms differ slightly in the chemical groups (represented here by R's) attached to the outside of the porphyrin ring. Most of them are small groups of atoms such as CH_3 and COOH, but R_7 is always a long chain of carbon atoms—as many as twenty—and hydrogen atoms. This "tail" is thought to embed the molecule in thylakoid membranes (see Figure 2-13).

Photosynthesis

The evolution of photosynthesis can be partly reconstructed by studying the bacterial forms that have not died out. Anaerobic photosynthetic bacteria lurking in oxygen-depleted muds and salt flats today are survivors from the variety that must have been more widespread in the early days. The three modern types are the green sulfur bacteria, the purple sulfur bacteria, and the purple nonsulfur bacteria. (The significance of these names will be explained.) These bacteria contain two kinds of chlorophyll: chlorobium (the green sulfur bacteria) and bacteriochlorophyll (both types of purple bacteria). In addition, they contain carotenoids, red and yellow pigments that function as accessories—they can absorb light of wavelengths that are not absorbed by chlorophyll, and pass the energy on to chlorophyll. The various pigments and other lipids, embedded in proteins, comprise photosynthetic membranes called thylakoids (see Figure 2-13).

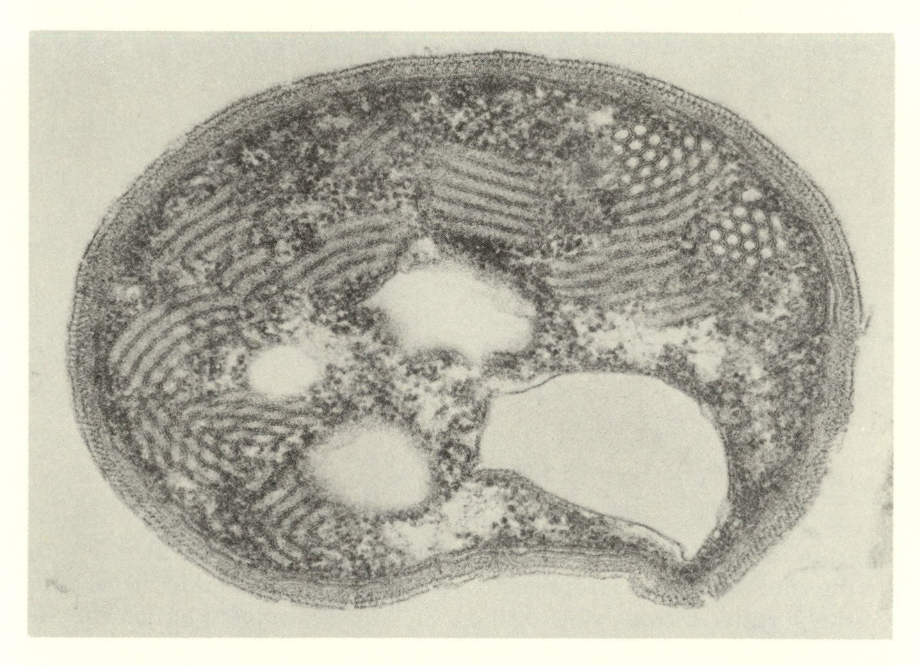

Figure 2-13. A purple sulfur photosynthetic bacterium, probably of the genus *Thiocapsa*. The thylakoid membranes are visible, in cross section, as clusters of short parallel lines. The cell is about 2 micrometers (millionths of a meter) long. [Courtesy of David Chase, Sepulveda Veterans Administration Hospital, Sepulveda, California.]

In all photosynthetic organisms, two separate sets of chemical reactions make up the photosynthetic process. Only those of the first set, the "light reactions," use light directly. When any molecule absorbs light, its electrons are boosted to a higher energy state. What happens to this energy? In most cases, the excited molecule emits the energy as light or heat, thereby returning to its normal, unexcited state. In fact, chlorophyll molecules not in a cell (say, dissolved in alcohol) do just that—they fluoresce for a while after they have been exposed to light. In photosynthetic cells, however, whose chlorophyll molecules are packed together in units, an excited molecule will pass its energy to an adjacent one. This continues at random until the energy reaches a special chlorophyll molecule (one in each unit) called the reaction center. Because the pigment molecules are neatly aligned with electron-transport chains in the thylakoids, high-energy electrons pass easily from excited reaction centers

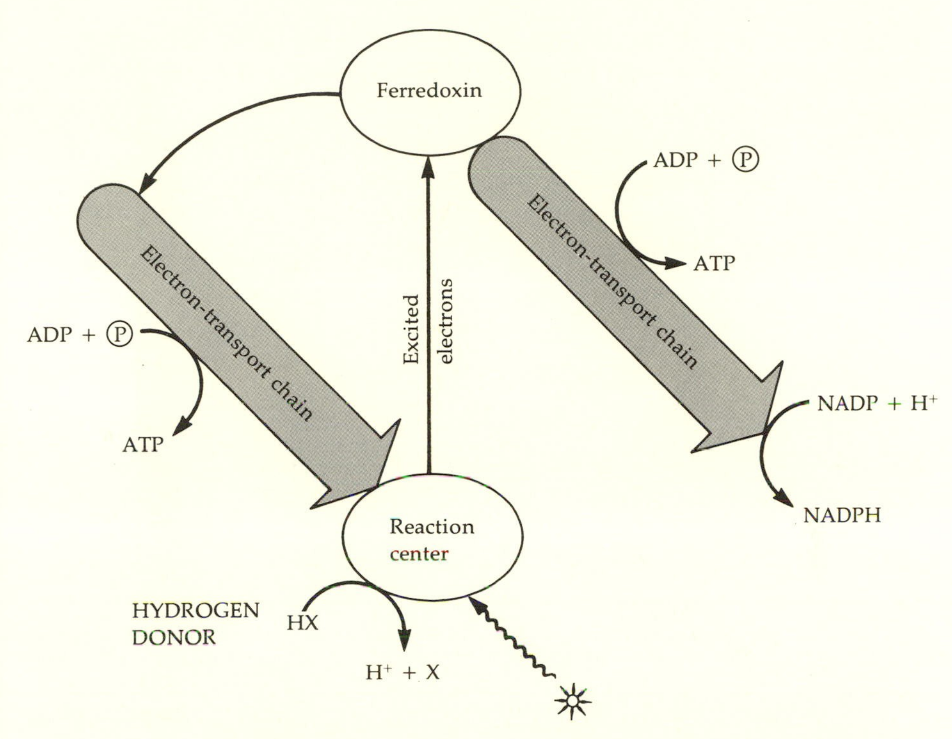

Figure 2-14. Diagram of the pathways of light-excited electrons in an anaerobic photosynthetic bacterium. Some of the electrons enter an electron-transport chain that produces ATP and delivers electrons back to the reaction center. This pathway is called cyclic photophosphorylation. The other pathway, on the right, delivers the electron to an NADP molecule, which then combines with a hydrogen ion (H^+, always present in water) to make NADPH. The reaction center resupplies itself with electrons, and its watery surroundings with hydrogen ions, by splitting a hydrogen compound (HX). Depending on the organism, HX can be hydrogen sulfide (H_2S), an organic compound, or even hydrogen gas (H_2).

into the chains, which generate ATP (see Figure 2-14).

Electron-transport chains in a photosynthetic cell are of two kinds. In one, the "used" electrons, have given up their energy to form ATP, are simply returned to an oxidized chlorophyll molecule. In the other, more frequently used pathway, not all the energy in the excited electrons is used to form ATP. Still energetic electrons are passed to the molecule NADP (nicotinamide adenine dinucleotide phosphate), which is thus reduced to NADPH. This reduced organic compound serves as a source of energy and hydrogen atoms in the second set of photosynthetic reactions.

Once the light reactions have built up stores of ATP and NADPH, the second set of reactions, or so-called "dark reactions," can carry on independently of sunshine, reducing carbon dioxide to organic compounds such as glucose:

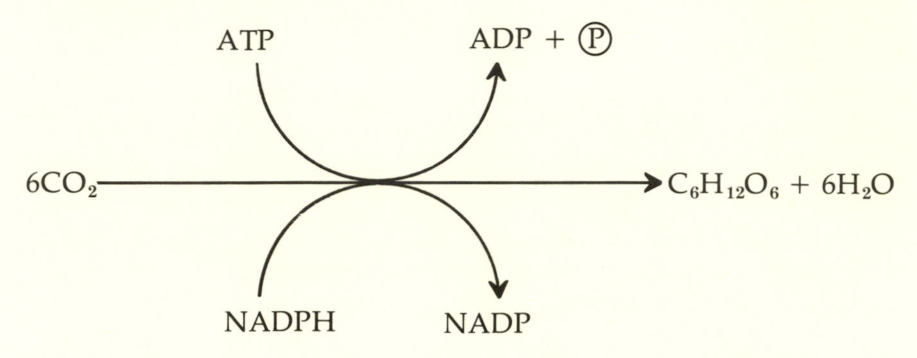

Cells either use the carbon compounds (photosynthate) generated during the dark reactions to build up cell material or they store them for future use. With development of the ability to convert atmospheric carbon dioxide into stored food, a characteristic of all photosynthesizers—from purple sulfur bacteria to today's giant redwoods—the Archaean food crisis subsided. In filling their own bodies with food reserves, photosynthesizers provided food for the heterotrophs that depended on them. It is likely that communities of photosynthetic microbes and their heterotrophic dependents thrived throughout the Archaean. The impressive amount of carbon-rich shales and cherts—probably residues of photosynthesizers—found in southern Africa by Barghoorn and the West German geologist Thomas Reimer indicate that such communities abounded as long ago as 3.4 billion years. Some rocks of this age have seams of carbon so extensive that they resemble the coal seams derived from tropical forests three billion years later.

For photosynthesis to take place, cells must have a source of hydrogen atoms to reduce carbon dioxide. The green and purple sulfur bacteria take their hydrogen atoms from volcanically or biologically produced hydrogen sulfide and generate sulfur as a waste product (hence their name). The nonsulfur purple bacteria remove hydrogen atoms from organic compounds such as ethanol, lactic acid, or pyruvic acid, or from hydrogen gas itself (H_2). Hydrogen atoms are cleaved from these sources when a reaction center absorbs light energy.

Hydrogen sources can be provided by neighboring bacteria as the end products of their fermentation processes. These cycles still take place, in anaerobic, light-filled muds, soils, and salt flats. In the anaerobic Archaean, hydrogen sulfide and hydrogen-containing compounds would have been more stable and abundant than they are today, when they are rapidly oxidized to water, carbon dioxide, and sulfur oxides by the oxygen in the air. The ultimate achievement, however, would be to use the hydrogen of water, a far more abundant and stable resource than organic compounds or hydrogen sulfide. When cyanobacteria evolved the capacity to split the water molecule for its hydrogen atoms, as I shall describe in the next chapter, they became a spectacularly successful group.

New Dangers

Photosynthesizing organisms require light; therefore, early photosynthesizing cells must have quickly developed the capacity to sense lighted areas and move toward them. In fact, all motile microbes swim in directed ways, not only toward light, but also toward food and away from acids, poisons, and other insults from their environment. It seems probable then that the combination of motility with simple chemical sensory systems arose in extremely ancient times. In the anaerobic muds of the Archaean, heterotrophic microbes that could chase their food must have had a great advantage over their immobile competitors. Later, anaerobic photosynthesizing cells could seek out better-lit places, thereby enhancing their photosynthetic production.

But there was a rub. With the visible light that bacteria needed came the stronger and more dangerous ultraviolet light. A cell's nucleic acids and proteins strongly absorb ultraviolet light, which ruptures chemical bonds and eventually causes the death of the cell. Nonphotosynthetic organisms could avoid this problem simply by seeking darkness, but this solution was not available to the photosynthesizers.

Several mechanisms of protection against ultraviolet radiation have been found in extant organisms, some presumably relicts from a time when ultraviolet radiation constituted a greater threat. Some microbes have developed protective mechanisms analogous to sunglasses: they live in solutions rich in certain salts, such as sodium nitrate, or under sand or other substances that absorb ultraviolet but let in visible light. Others have developed protection mechanisms analogous to suntanning: they synthesize pigments that absorb ultraviolet radiation.

Finally, some cells live in matlike communities, which protect them from unwanted solar radiation. The cells on top die, most likely from overexposure to solar radiation, but their corpses then protect the lower layers of microbes. Many such sedimentary structures, called stromatolites, persist from before the Cambrian. As I shall describe in the next chapter, stromatolites are important sources of evidence of Archaean life.

Cells not only protect themselves against radiation damage to their nucleic acids, they have also evolved ways of repairing such damage. No doubt early in the history of life, cells developed certain DNA polymerases and other enzymes that enabled them to repair their damaged genetic material in the same way that DNA is replicated for cell reproduction—by "reading" the sequence of nucleotides of one strand to sense and repair defects in the sequence of the complementary strand. In some organisms, these enzymes must be activated by visible light. If, immediately after ultraviolet irradiation, these organisms are left in darkness, no patching of their DNA takes place, whereas if they receive strong visible light immediately afterward, only minimal damage remains.

Repair enzymes must have been indispensable to light-requiring photosynthetic bacteria in the Archaean. The absence of free oxygen meant that there was no ozone layer high in the atmosphere to block the penetration of short-wavelength, DNA-damaging ultraviolet light. Today, only the longest-wavelength ultraviolet and visible light can enter freely, yet nearly all organisms still have enzymes for repairing ultraviolet damage to DNA. Two billion years after the formation of the ozone layer, why should these enzymes, which have no function in DNA replication, persist? This sophisticated method of handling dangerous ultraviolet light probably has been retained because it made possible another evolutionary innovation—first sex.

First Sex

All the bacteria that I have described reproduce directly by division. The parent cell makes a new copy of its single DNA molecule, which, like the old DNA, attaches itself to the inner cell membrane. As the membrane grows, the new and old DNA molecules separate into two genophores ("gene bearers"). Eventually, the cell synthesizes a new cell wall that divides it in half and the two cells separate by fission—direct and equal cell division.

Some bacteria reproduce by budding. The parent cell produces a small attached offspring cell that contains a complete supply of cell material, including a copy of the parent's genes. The offspring gradually grows until it reaches parent size, when it breaks off.

Finally, certain bacteria have developed the capacity to form an internal spore whose walls envelop the parent cell's genetic material. The walls of the spore undergo a chemical alteration that enables them to tolerate long periods of desiccation, heat, and the absence of food. Once the alteration is complete, the cell disintegrates, releasing the spore, which eventually germinates into a new cell when conditions become favorable.

Bacteria that reproduce in these ways have only one parent and receive all their genes from that parent. Single-parent reproduction is called asexual. Some bacteria have also developed a kind of sexual process called recombination or bacterial conjugation, in which the offspring contains genes from two parents. When two bacteria of compatible strains happen to be near each other, a thin bridge of cell material forms between them. Through this bridge, one of the bacteria, called the donor, transfers some of its DNA to the other, called the recipient (see Figure 2-15). The amount of DNA that is transferred is extremely variable, ranging from as little as one gene to as much as an entire set of the donor's genes. By definition, or, perhaps better said, by analogy, the donor is considered male, the recipient female. After conjugation, the recipient, now the offspring, is called the recombinant. In some cases, the donor, having given away its

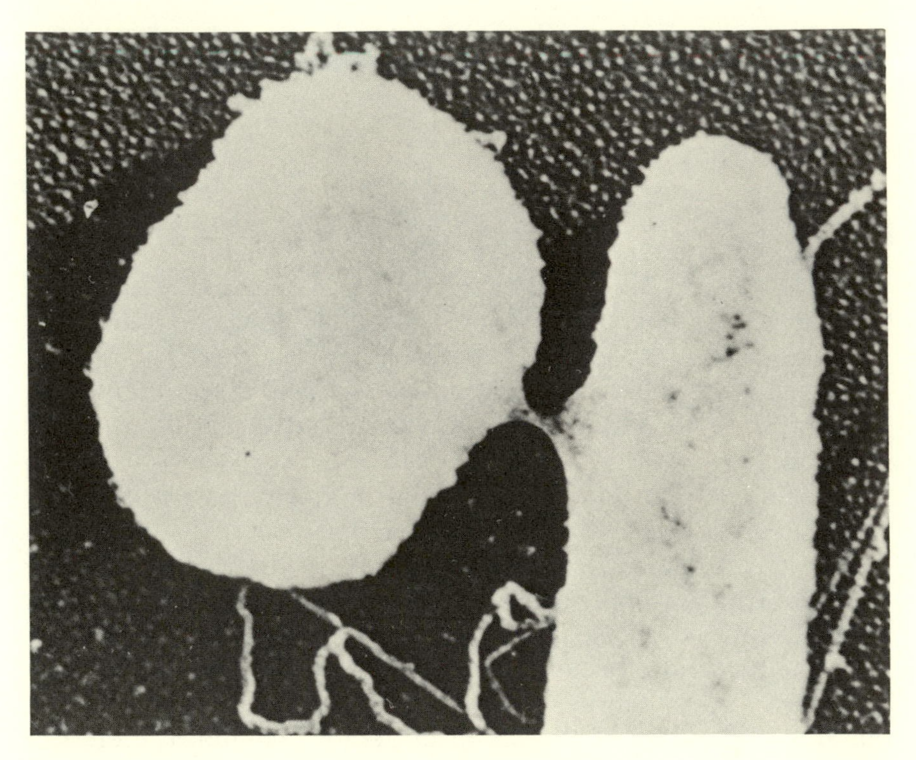

Figure 2-15. Conjugation between a donor ("male") and a recipient ("female") *Escherichia coli*, a bacterium common in the human intestine. The long, thin partner is the recipient. [Courtesy of T.F. Anderson, E.L. Wollman, and F. Jacob, Institute for Cancer Research, Philadelphia.]

genes, then dies. Unlike the offspring of animals and plants, recombinant bacteria rarely have received exactly half of their genes from each parent.

After conjugation, the pieces of DNA from the two parents are spliced together to form a new long DNA molecule. This splicing requires some of the same enzymes that are used to patch ultraviolet-damaged DNA. Evidently, sex in this limited sense was able to arise because there was already a way for broken strands of DNA to recombine after they were damaged by ultraviolet irradiation. Recombination offered populations of bacteria a means of rapidly responding to changing (and therefore threatening) environmental conditions: adaptive genes could pass from one bacterium to another with ease. It was no longer necessary to wait for favorable and permanent mutations to occur in a single organism and its direct descendants.

Ecology without Oxygen

Changes in the environment caused life to change, but the reverse was equally true. Once life had appeared on Earth, the planet was never again the same. Living microbes interacted with the surface sediments and the atmosphere, hastening and modulating the major cycles of chemical elements. A striking difference between the Earth and its two planetary neighbors, for example, is the lower proportion of carbon dioxide in the Earth's atmosphere (see Table 2-1). At least some of that depletion resulted from the early removal of carbon dioxide from the air and its transfer into the sediments by such anaerobic microbes as the clostridia, desulfovibrios, and anaerobic photosynthesizers all of which fix carbon dioxide.

Table 2-1. Percentages of Some Gases in the Atmospheres of Venus, Earth, and Mars

	Venus	Earth	Mars
Carbon dioxide	98	0.03	95
Nitrogen	1.9	79	2.7
Oxygen	Trace	21	0.1

From the known metabolic properties of anaerobic cells that have survived until the present time, biologists have been able to sketch a tentative picture of microbial gas exchange in the Archaean (see Figure 2-16). Cells required (as they still require) the following biologically critical elements: hydrogen, carbon, nitrogen, oxygen, phosphorus, and sulfur. These elements had to reach organisms in chemically available form; those that were tied up in rocks, as carbon is in shales for example, would have been of no use to the biota. To be useful, they had to be transported by the fluids of the Earth—carried by the atmosphere or hydrosphere. Archaean volcanoes spewed hydrogen and nitro-

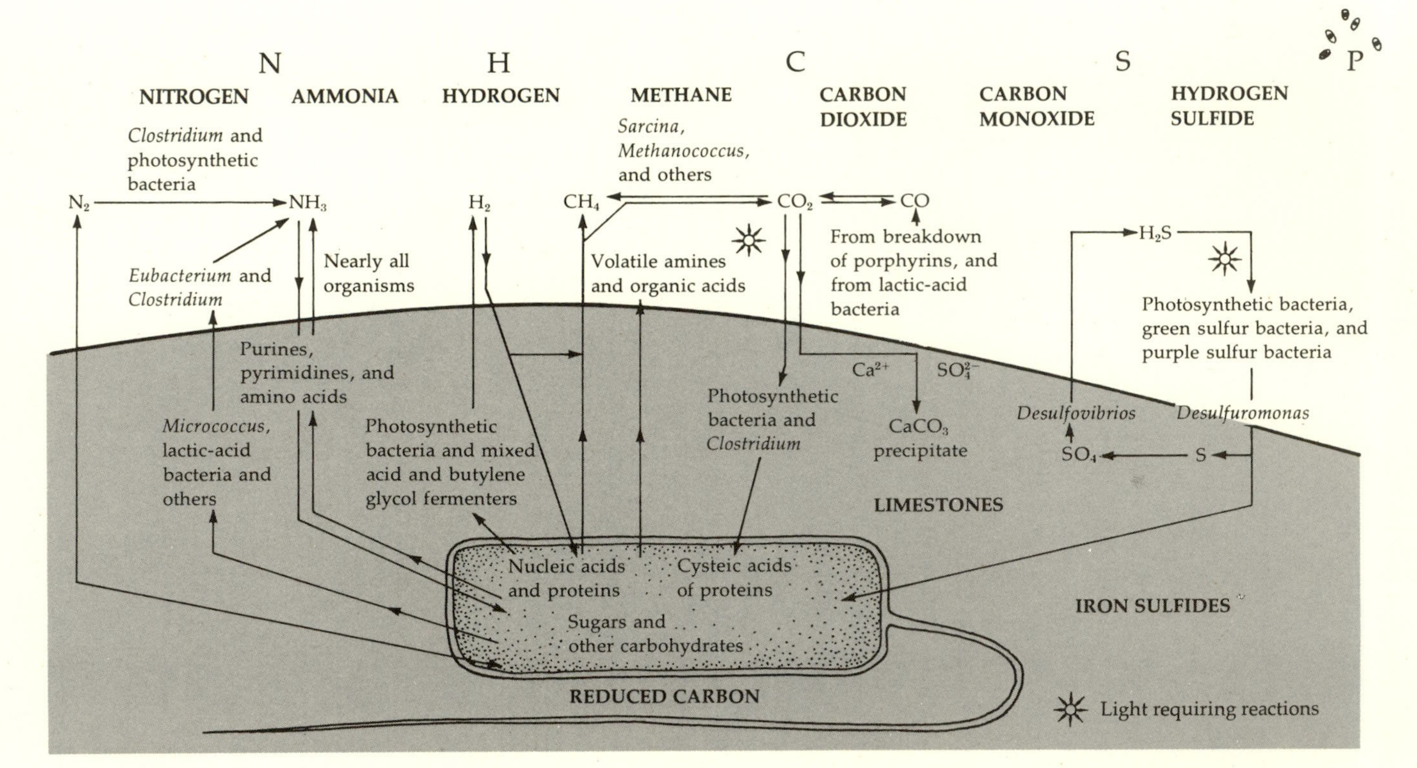

Figure 2-16. A model of the cycling of essential elements in a world without free oxygen. The dark curved line crossing the diagram represents a horizon—above is the atmosphere, below is water and sediment. Because phosphorus has no gaseous phase, except perhaps phosphine (PH_3), it may have been spread through the atmosphere in spores and other biological particles. [From L. Margulis, *Symbiosis in Cell Evolution*. San Francisco: W.H. Freeman and Co., 1981.]

gen gases into the atmosphere, and many early cells could take up these elements directly from the air, which also circulated carbon dioxide, methane, and hydrogen sulfide gases. Elements circulated in water-soluble compounds such as ammonium (NH_4^+), nitrates (NO_3^-), nitrites (NO_2^-), carbon dioxide, sulfates, and phosphates. Phosphorus also circulated in airborne particles, such as spores. Even today, some of these cycles, such as the nitrogen cycle, can be completed only by anaerobic bacteria.

Today's oxygen-intolerant microbes are like lost tribes of people in remote areas, out of contact with the rest of the world. They have developed responses to geologic and chemical conditions that probably have not existed on a worldwide scale since the beginning of the Archaean. The organisms that succeeded them—as the next chapter explains—added the powerfully reactive gas oxygen to the major Earth–life cycles, and anaerobic cells had to retreat to mud flats and other airless nooks, where they still provide critical links in the cycling of gases and soluble compounds through the atmosphere and waters. They may be relicts from the earliest days of life, but we, the rest of life, would not survive if they were to become extinct.

Suggested Reading

Brock, T. D. *Biology of Microorganisms,* 3rd ed. Englewood Cliffs, N. J.: Prentice-Hall, 1979.

Day, W. *Genesis on Planet Earth.* E. Lansing, Mich.: House of Talos, 1979.

Haldane, J. B. S. *The Origin of Life.* 1929. [Reprinted in J. D. Bernal, *The Origin of Life.* Cleveland: World, 1967.]

Oparin, A. I. *The Origin of Life.* 1924. [Translated from Russian and reprinted in J. D. Bernal, *The Origin of Life.* Cleveland: World, 1967.]

Sieburth, J. M. *Sea Microbes.* New York: Oxford Univ. Press, 1979.

CHAPTER 3

Life With Oxygen

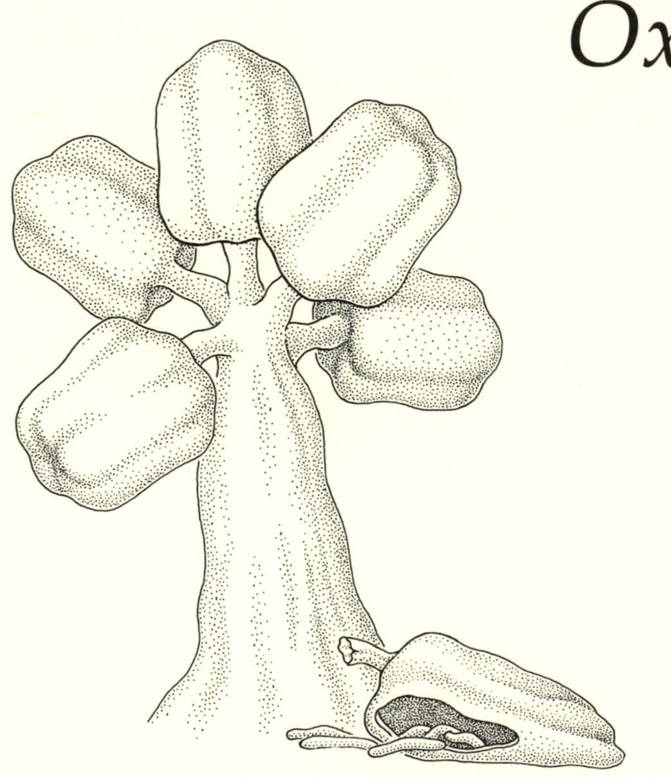

WHEN LIFE began in the Archaean seas, cells could simply take up surrounding organic compounds and use them directly for growth and reproduction. As these prebiotic organic compounds became depleted, anaerobic bacterial photosynthesis evolved and life gained its independence from preformed food. Microbes diversified: in many low-lying mud communities, cells carried out numerous metabolic strategies, including fermentation, sulfate reduction, and anaerobic photosynthesis.

Another shortage, this time of reducing agents, then threatened life's existence. The hydrogen sources on which photosynthesizing cells depended—hydrogen gas, hydrogen sulfide, and small hydrogen-containing organic molecules—became less plentiful, as hydrogen itself was constantly escaping into space. In response, cells evolved a new kind of photosynthesis that enabled them to remove hydrogen atoms from water, a virtually limitless source. In splitting the water molecule, photosynthesizers took the hydrogen atoms they needed and then eliminated oxygen as waste. Gradually, oxygen began to accumulate in water, soils, sediments, and the atmosphere. The aerobic era began.

Oxygen, a reactive, toxic gas that can oxidize carbon compounds, was a mixed blessing. Its appearance in the atmosphere precluded the further nonbiological synthesis of organic compounds. Not only did free oxygen quickly oxidize such compounds into carbon dioxide and water, but its presence led to the formation of the ozone layer, which greatly reduced the amount of solar ultraviolet radiation reaching the Earth's surface. Many organisms, unable to tolerate the gas, were permanently driven to anaerobic niches. However, oxygen also made possible a new energy-generating pathway—aerobic respiration. The greater efficiency of this pathway then provided cells with enough energy to grow larger and become structurally more elaborate. Respiration was prerequisite to the origin of the eukaryotic cells that became the building blocks of larger organisms.

The First Oxygen Producers

Probably the first organisms to generate oxygen photosynthetically were the cyanobacteria, traditionally called "blue-green algae." Long before the Cambrian, these cells had their days of abundance, diversity, and domination of the landscape. They first appeared in the Archaean; by the late Proterozoic, two billion years later, many thousands of species had spread over the Earth. In the Proterozoic, they built major rock formations, like extensive reefs. These became less abundant as the Proterozoic drew to a close. Today there are still several thousand species of cyanobacteria. They form scums on ponds, puddles, rice paddies, and shower curtains—wherever they can find light, water, and some security from scavengers. Their decline from dominance was probably due to many factors, not the least of which was the immense success of the eukaryotic algae and plants that succeeded them.

Except for their ability to split water, cyanobacteria are like anaerobic sulfur photosynthetic bacteria in nearly every way: they have the same type of cell structure, cell walls, and membranes (see Figure 3-1). Some cyanobacteria and anaerobic photosynthetic prokaryotes have gas vacuoles, membrane-bound gas sacs in the cell cytoplasm. Gas vacuoles regulate buoyancy, an important characteristic for organisms that live in water and require sunlight. Gas vacuoles enable such organisms to regulate their depth: by letting gas out, they can lower themselves to obtain nutrients or to escape oxygen gas at the surface; by filling their vacuoles with gas, they can raise themselves toward sunlight.

That cyanobacteria evolved from anaerobic photosynthetic bacteria had been suspected since the nineteenth century, when they were first observed clearly under the microscope. Now much evidence supports the conjecture. In 1975, a group of Israeli microbiologists, working in Solar Lake in the Negev Desert, a

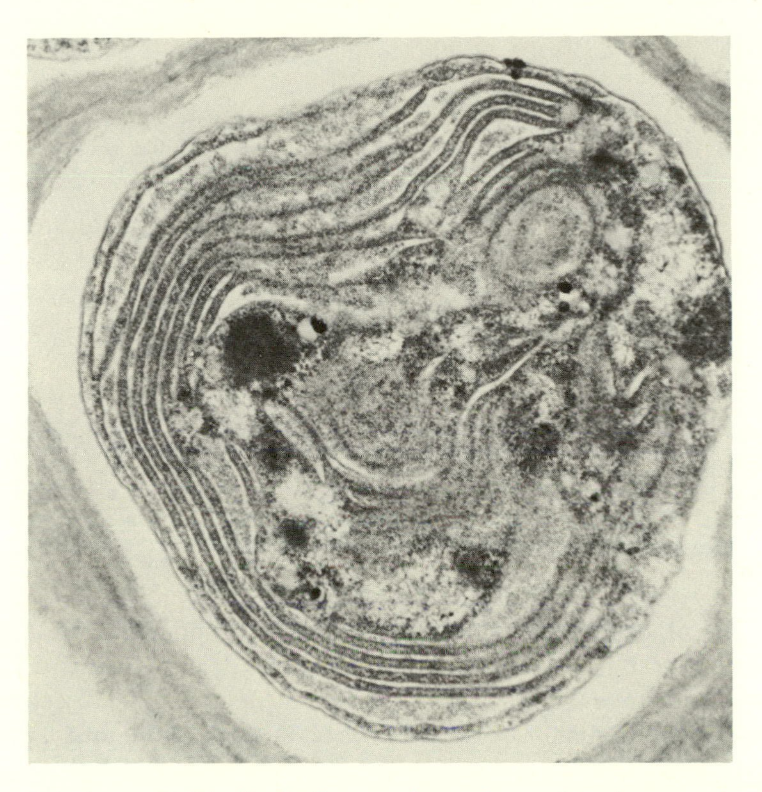

Figure 3-1. A coccoid (spherical) cyanobacterium taken from a microbial mat at Laguna Figueroa, Baja California. The parallel membranes clearly visible near the edge of the cell are thylakoids. Compare them with the thylakoids of the purple sulfur bacterium shown in Figure 2-13. The cell is about 5 micrometers (millionths of a meter) wide. [Courtesy of John Stolz, Boston University.]

small artificial pond in which the hottest water is near the bottom, found the filamentous organism *Oscillatoria limnetica* (see Figure 3-2). This beautiful cyanobacterium, named for its gliding movements, turned out to be a living fossil, a missing link between the anaerobic sulfur photosynthetic bacteria and the cyanobacteria. If placed in an environment where oxygen concentration is low and hydrogen sulfide concentration is high, it acts like an anaerobic photosynthetic bacterium: instead of using water as its hydrogen donor in photosynthesis, it reverts to the ancestral mode and uses hydrogen sulfide. After the discovery of *O. limnetica,* researchers quickly learned that a variety of cyanobacteria can revert to "anaerobic style" photosynthesis if

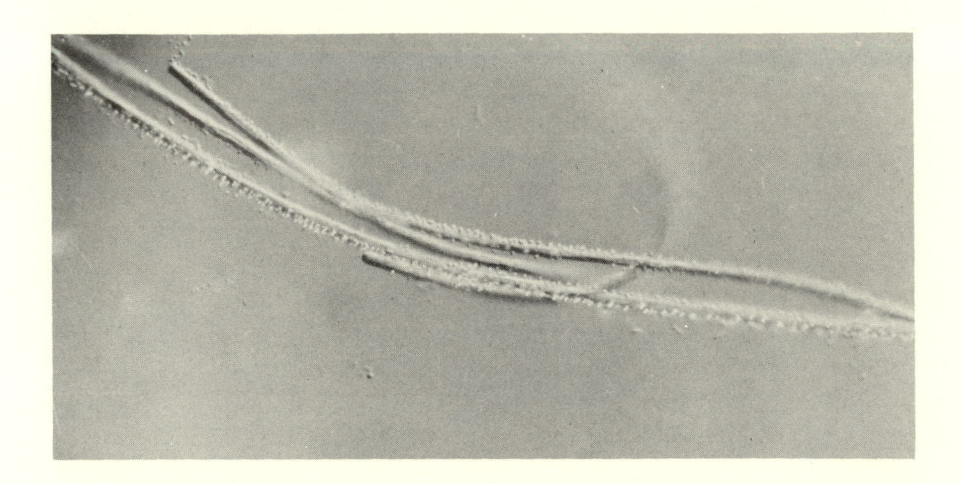

Figure 3-2. Oscillatoria limnetica. [Courtesy of Yehuda Cohen, Eilat Marine Station, Eilat, Israel.]

they are placed in suitable conditions. One suspects that such jack-of-all-trades photosynthesizers abounded before the Cambrian, but have subsequently been squeezed out by photosynthetic specialists, both anaerobic sulfur users and aerobic water users. They may now be limited to certain rather extreme environments.

Cyanobacteria do not contain the bacteriochlorophylls or chlorobium chlorophylls found in anaerobic photosynthesizers, but do contain chlorophyll *a*, one of the pigments found in algae and plants. Like the chlorophyll of anaerobic photosynthesizers, chlorophyll *a* along with certain other lipids and proteins comprises thylakoid membranes in the cyanobacterial cell. Unlike anaerobic photosynthesizers, cyanobacteria have two kinds of reaction centers, and this is what enables them to break down water for its hydrogen.

The reaction centers of one kind, and the electron-transport chain that they feed into, are virtually identical to the system of the anaerobic photosynthesizers (see Figure 2-14). This system, called photosystem I, is activated by light of a certain wavelength. In the anaerobic photosynthesizers, the energy in the light is used not only to excite electrons, but also to split molecules for their hydrogen atoms. However, the hydrogen atoms in a water molecule are bound very firmly to the oxygen atom. Splitting water takes more energy than splitting hydrogen

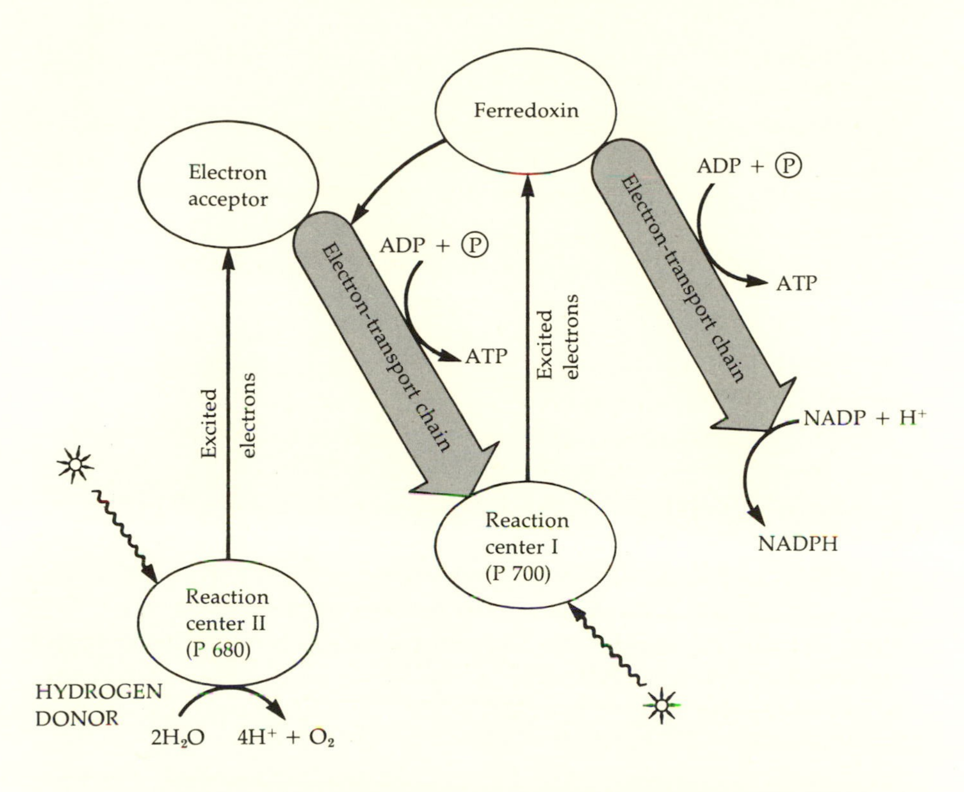

Figure 3-3. The pathway of light-excited electrons in cyanobacteria and plants. The ultimate fate of the electrons is to combine with NADP and hydrogen ions (H^+) to produce NADPH. The source of the electrons, and of hydrogen ions needed to replace those removed from the watery medium, is water molecules split by reaction center II. P680 and P700 stand for chlorophyll pigments that absorb light at wavelengths of 680 and 700 nanometers (billionths of a meter). The pathway from reaction center I onward is identical to the photosynthetic pathway in anaerobic photosynthesizers (see Figure 2-14).

sulfide, hydrogen gas, or organic molecules, and the energy of the light waves absorbed by photosystem I is insufficient.

The innovation of the cyanobacteria was a new kind of reaction center with its associated proteins and enzymes, called photosystem II. This system absorbs light of a shorter wavelength (and therefore is more energetic). Photosystem II splits water molecules and sends excited electrons into an ATP-producing electron-transport chain. The "used" electrons do not return to the reaction center that produced them, however, but to the reaction centers of photosystem I (see Figure 3-3). There they are re-excited by absorbed light and used to produce NADPH and more ATP.

All plants, algae, and cyanobacteria—all oxygen-releasing photosynthesizers—contain these two linked photosystems; no organism is known to contain only photosystem II, although all the anaerobic photosynthesizers contain only photosystem I. This suggests that photosystem I evolved first and that oxygenic photosynthesis evolved only after photosystem I was working well. The combined photosystems generate more molecules of ATP than photosystem I alone, and thus cyanobacteria are more efficient photosynthesizers than their anaerobic ancestors. Their efficiency and their unique access to the most abundant hydrogen source on the planet were to make the cyanobacteria spectacularly successful. For the history of life on our planet, however, the most important feature of cyanobacterial photosynthesis is almost incidental: the waste product. Having combined the hydrogen atoms from water with carbon dioxide to form organic compounds, cyanobacteria then eliminated the residual oxygen molecules. The accumulation of the waste of these microbes was the start of the oxygen revolution.

The Transformation of the Atmosphere

The rise of aerobic photosynthesis was a global catastrophe. Oxygen "grabs" electrons and produces so-called free radicals: highly reactive, short-lived chemical species that wreak havoc with organic compounds. Fermenting bacteria and photosynthetic sulfur bacteria retreated to zones where oxygen is scarce. Today, hundreds of obligately anaerobic species are limited to niches in decaying swamp vegetation, mud flats, and lake bottoms. Some fermenting bacteria even colonize anaerobic zones of the digestive tracts of animals, such as termites and ruminants, niches that became available only after animals evolved, in the Phanerozoic. (Cows, in fact, have been aptly described as 40-gallon methane-fermenting tanks standing on four legs.)

At first, as oxygen came from the cyanobacteria, it must have combined quickly with other gases, including hydrogen, ammonia, carbon monoxide, and hydrogen sulfide. Nowadays, hydrogen sulfide produced by lake-bottom bacteria combines with oxygen before it reaches the surface of the water; hydrogen sulfide, therefore, cannot accumulate. Likewise, the oxygen produced by the first cyanobacteria did not accumulate in the atmosphere as long as there were abundant gases that could react with it.

Oxygen also combined with many soil minerals dissolved in oceans, rivers, and lakes. Iron, for example, reacts with oxygen easily. Evidence that such reactions occurred persists in the geological record. Many sedimentary rocks are formed by processes that involve reworking and transporting surface materials, which are thus brought into extensive contact with the atmosphere. If such rocks form in an atmosphere containing abundant oxygen, many of the minerals in the rocks react with the oxygen to form their oxidized counterparts. The presence of oxidized forms of iron, manganese, uranium, and other elements in rocks from the Proterozoic, two billion years ago, argues strongly that the atmosphere has been oxygenic since at least that time.

In contrast, the minerals found in the older, Archaean rocks suggest that the surface of the Earth was then lacking in oxygen. Pebbles that rolled in Archaean rivers or streams and were then buried show no evidence of having contacted oxygen. They contain oxygen-poor forms of sulfur, uranium, and iron. Iron tends to be in the reduced, or less oxidized (Fe^{2+}, ferrous) form, rather than in the more oxidized (Fe^{3+}, ferric) form. Similarly, the less oxidized form of uranium (uraninite) is found only in sediments more than two billion years old; thereafter, oxidized forms became common. Sulfur, too, is found in Archaean rocks in the less oxidized forms (sulfides), whereas the oxidized forms (sulfates) are rare. In the iron-rich Michepecoten–Woman River formation (more than 2.5 billion years old) of western Ontario, for example, Henry Thode and his colleagues from MacMaster University have found evidence of a large, open body of water, an ancient anaerobic sea in which desulfovibrios dominated the surface of the shoreline. Because desulfovibrios cannot tolerate oxygen, they are now always found below the surface. In Archaean times, these microbes had reduced sulfate (from the seawater) to sulfide anaerobically over a large area, leaving large sulfide deposits, testimony that more than two billion years ago the concentration of oxygen in the air was still low.

Much of the evidence concerning the transition to the oxidizing atmosphere comes from studies of geologic features known as banded iron formations (see Figure 3-4). These sediments cover great expanses of Australia, North America, and Africa. The formations date from Archaean times and are made largely of layers containing more reduced iron (magnetite) alternating with layers containing more oxidized iron (hematite). One interpretation of such features is that the sediments were laid down under varying concentrations of atmospheric oxygen. When oxygen was absent, ferrous (reduced) iron was transported and sedimented; it is darker in color than the rustlike ferric (oxidized) iron that was deposited when oxygen was present. In rocks laid down after about two billion years ago, however, nearly all the iron deposited has the fully-oxidized, rusty appearance characteristic of sediments formed in the presence of oxygen. About 1.8 billion years ago, large banded iron formations abruptly and entirely disappeared, to be replaced by sediments known as red beds. This suggests that by this time the Earth's atmosphere had accumulated a significant concentration of oxygen.

Not all investigators agree with this interpretation of banded iron formations. Preston Cloud of the University of California at Santa Barbara and his co-worker Aharon Gibor believe that, about two billion years ago, the atmospheric concentration of oxygen was less than one percent of the present concentration. Even 600 million years ago, they believe, when well-developed animals were already abundant, the concentration of atmospheric oxygen was still only a few percent of the present value. In their opinion, the presence of dangerous ultraviolet radiation and the lack of oxygen itself explains the late appearance of animals.

I disagree. It is my opinion that two billion years ago the concentration of oxygen and the density of the ozone screen were at least 50 percent of their modern values, and that 600 million years ago atmospheric oxygen was already at (or nearly at) its present concentration. James C. G. Walker of the University of Michigan, Mitchell B. Rambler, of Boston University, and I have studied the effect of ultraviolet rays on anaerobic bacteria (*Clostridium* as well as some that can grow either with or without oxygen) and on mixed communities of mat-forming microbes. We were surprised to find a great degree of resistance by such organisms to large doses of ultraviolet—even larger than the dose corresponding to full sunlight in the absence of ozone. We see no

Figure 3-4. A sample of a metamorphosed banded iron formation about two billion years old from Negaunee Iron Formation, Marquette, Michigan. The darker bands (black) contain specular hematite; the lighter ones (reddish) contain jaspilite, chert with finely dispersed hematite. Both are in a chert matrix. (About half natural size.) [Courtesy of Harold L. James, U.S.G.S., Port Townsend, Washington.]

evidence that ultraviolet itself ever prevented the colonization of the land or the evolution of multicellular organisms. Furthermore, Walker's calculations point to the conclusion that once oxygen began to accumulate in the atmosphere, it did so rapidly and reached about the present concentration by about 1.8 billion years ago.

Although photosynthesizers began generating oxygen earlier in the Archaean, for millions of years other molecules—reduced sulfur, reduced carbon, and volcanic gases such as hydrogen—constituted a "sink" for the oxygen. That is, the oxidation of these molecules prevented any net accumulation of oxygen in the atmosphere. Robert Garrels, a geologist at the University of Southern Florida (St. Petersburg), believes that reduced sulfur was the major sink for oxygen and that only when most of the sulfur was oxidized did oxygen begin to accumulate. In any case, once oxygen began to accumulate, it did so rapidly because there was nothing to prevent it.

Stromatolites

Although the chronological details are in question, everyone who has studied the evidence agrees that a transition occurred from an anaerobic to an aerobic world. It is also accepted that the concentration of oxygen in the atmosphere was never much greater than the modern value—20 percent. If it had been, the fossil record would contain evidence of worldwide conflagration, because even rain forest and damp grasslands are extremely flammable in the presence of high oxygen concentrations. Spontaneous combustion of microorganisms would also have occurred. Although many local fossil fires have been detected, there is no evidence for worldwide conflagration.

The geological record also contains good evidence bearing on the activities and spread of cyanobacteria themselves. The evidence is in the form of stromatolites, dome-shaped, layered rocks present throughout the fossil record, but especially conspicuous in rocks from before the Cambrian (see Figure 3-5). These sedimentary structures are usually composed of calcium carbonate (the predominant compound in chalk and limestone), although very few, such as those in the Gunflint Iron Formation, western Ontario, are made primarily of chert (silicon dioxide). Some stromatolites are conical or cauliflower shaped, some are columnar, others are flat. Fields of them can extend over hectares of surface or may be limited to a few square meters.

Until recently, stromatolites were the object of some geological speculation, but little understanding. They were described systematically late in the nineteenth century, particularly by the

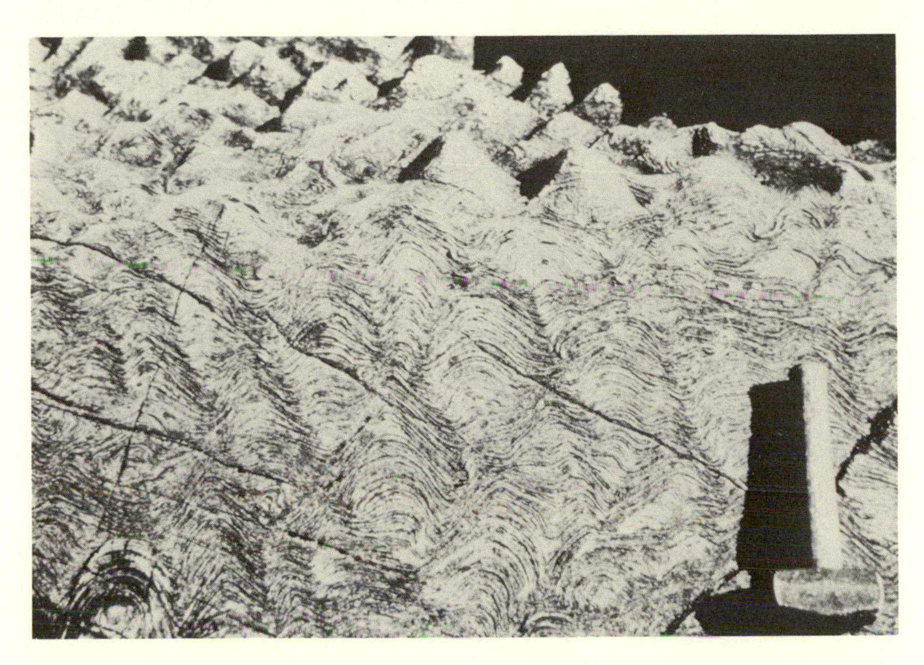

Figure 3-5. A cross section of some stromatolites more than two billion years old near Great Slave Lake, Canada. [Courtesy of Stanley M. Awramik, University of California at Santa Barbara.]

American geologist Charles Walcott, and they were given genus and species names, but investigators did not understand what the rocks represented. Some geologists, believing that stromatolites constituted the remains of living organisms, gave them the name *Cryptozoa* ("hidden animals"). In fact, these investigators were correct in interpreting stromatolites as fossils.

The precursor of a stromatolite is a microbial mat, an accreting surface or zone in which bacteria (primarily cyanobacteria) photosynthesize, grow, and reproduce. A typical mat looks like mere dirty sand or dark green scum, but one can see upon close inspection that it is fibrous and composed of intertwined filaments. Such a mat traps sediment—clay, mud particles, organic debris. As the sediment accumulates, the microbes either are trapped in it and die or move upward in the light to start a new layer atop the old. If the old layer contains enough sediment and the evaporation rate is high, it gradually turns to stone. As these events repeat themselves periodically, a many-layered stromato-

Figure 3-6. Modern stromatolites in Hamelin Pool, Shark Bay, Western Australia. Although the pool is connected with the sea, its water is too salty for the worms and mollusks that would graze on the mats of microbes and prevent them from forming stromatolites. [Courtesy of Stanley M. Awramik, University of California at Santa Barbara.]

lite takes form. Stromatolites are as stable as most other sedimentary rocks, although their structure reveals them to be the work of living things (see Figure 3-6).

Some ancient stromatolites contain fossilized microbes; their resemblance to modern forms of cyanobacteria is occasionally uncanny. The oldest known stromatolites are about 3.5 billion years old, from an area called North Pole in Western Australia. The structure of North Pole stromatolites and the microbial fossils (see Figure 2-6) in their carbon-rich layers are convincing evidence that photosynthetic microbial communities were thriving by that time.* The photosynthesizers may have been anaerobic photosynthetic bacteria, or, like *Oscillatoria limnetica*, optionally able to release oxygen, or they may have been entirely oxygen releasing. It is difficult to be sure. In any case, the first Archaean stromatolites probably formed before the ozone layer,

*See note on page 32.

so the stromatolitic way of life may have evolved partly to protect the microbes from lethal ultraviolet. Although many cells in the uppermost layer would die, their carcasses would filter out the destroying wavelengths to the benefit of cells deeper in the mat.

Learning to Breathe

As cyanobacteria spread over the surface of the Earth and the amount of atmospheric oxygen increased, all life forms were threatened, even the producers themselves. Even today, a few cyanobacteria, such as some species of *Phormidium,* are poisoned by the oxygen they produce; they live in anaerobic zones with aerobic microbes that quickly deplete the oxygen as it is generated. Other microbes, including certain cyanobacteria, can tolerate oxygen, but only in very low concentrations, and are happiest in oxygen-poor waters. Some microbes are indifferent; they neither seek nor shun oxygen, but simply tolerate it.

Organisms that can tolerate oxygen contain enzymes—catalases, peroxidases, and superoxide dismutases—that react with dangerous radicals produced by oxygen and convert them to innocuous organic compounds and water. The evolution of protective enzymes enabled cells to tolerate the presence of oxygen. Soon, however, microbes developed a solution to the oxygen problem that not only protected them from the potentially poisonous gas, but also provided them with another mode of energy transformation: they evolved the capacity to consume the oxygen they produced photosynthetically, thereby generating additional molecules of ATP. The photosynthate they produced by trapping the energy in light was oxidized inside the cell. In this way, photosynthesizing cells could produce far larger quantities of ATP than they had by fermentation, in the absence of oxygen. Aerobic respiration, the highly efficient oxidation of organic food molecules, had arrived.

Although it varies in detail among species of microbes, aerobic respiration generally can be viewed as a controlled combustion that breaks down organic molecules and yields carbon dioxide and water, energy-poor compounds. The released energy is harnessed to produce ATP. Typically, the fermentation of a sugar molecule produces two molecules of ATP; the respiration of the same sugar molecule can produce as many as 36. The ATP is generated in three stages (see Figure 3-7). The first is identical to

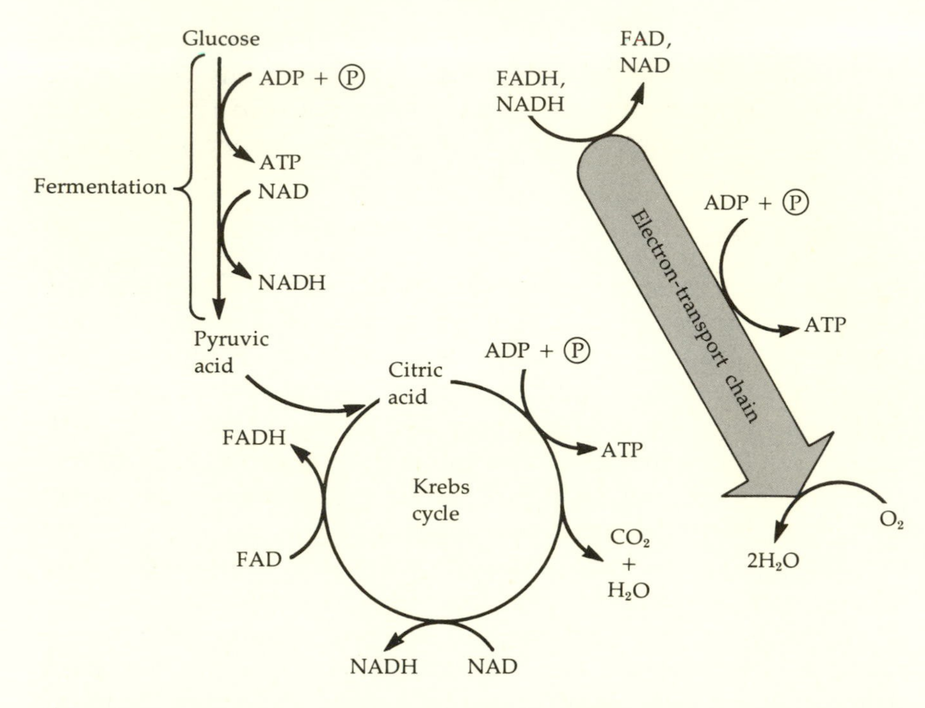

Figure 3-7. The three parts of aerobic respiration: the breakdown of glucose by fermentation, the Krebs cycle, and an electron-transport chain.

fermentation—a food molecule (say, glucose) is converted into pyruvic acid and some ATP is generated (see Figure 2-7).

The second stage is a cycle of reactions called the Krebs cycle (after its discoverer) or the citric acid cycle (after one of its components)—it consumes the pyruvic acid produced by the first stage of respiration. Some of the energy released is used to generate more ATP, but most of it is transferred as high-energy electrons to the carrier molecules FAD (flavin adenine dinucleotide) and NAD (nicotinamide adenine dinucleotide), reducing them to FADH and NADH. These molecules, which bear most of the energy from the original food molecule, transfer their electrons to the third stage of respiration, an electron-transport chain made up of several kinds of electron acceptors, including ferredoxins and cytochromes. This is the stage that produces most of the ATP—as many as 32 molecules of it from each food molecule. The final electron acceptor in the chain is molecular oxygen, which, upon receipt of electrons (and their attendant hydrogen atoms), is reduced to water. In all aerobically respiring organisms, the transfer of electrons to oxygen is the final step in the respiration pathway. In a breakdown pathway, this implies that it was the last step to evolve.

Some NADH is also produced in the first stage of respiration, and, in fact, fermenting bacteria produce it also. However, lacking an electron-transport chain, they must reconvert the NADH to NAD without deriving energy from it (see Figure 2-7). The sulfate-reducing bacteria did in a way foreshadow the appearance of aerobic metabolism—they feed high-energy electrons derived from a fermentation pathway into an electron-transport chain (see Figure 2-9). In those bacteria the final electron acceptor is sulfate, so they can be said to "breathe" sulfate by reducing it to hydrogen sulfide just as aerobic respirers breathe oxygen by reducing it to water.

Having made the oxygen with which they then had to cope, it was the cyanobacteria that probably were the first organisms to use it in respiration. They differ from other aerobic organisms in that they respire only in the dark. Why? Apparently, they use some of the same molecular machinery for their respiratory electron-transport chain and their photosynthetic electron-transport chain; the shared components cannot be used simultaneously for both pathways. Thus, because photosynthesis takes place in the light, respiration must take place in the dark. In algae and plants, which evolved later, the two processes can proceed at the same time because they take place in different parts of the cell: photosynthesis goes on in the chloroplasts while respiration consumes the excreted oxygen in the mitochondria.

With the refinement of oxygen-generating photosynthesis and oxygen-consuming respiration, cyanobacteria literally found their place in the sun. Given sunlight, a few salts always present in natural waters, and atmospheric carbon dioxide, they could make everything they needed: nucleic acids, proteins, and vitamins. The cyanobacteria diversified into hundreds of different forms. Some remained small, a fraction of a micrometer in diameter; other species grew large, as much as 80 micrometers in diameter. Some cyanobacteria took the form of simple spheres or of small cells embedded in a gelatinous matrix, whereas others developed into multicellular sheets of cells. Some evolved bundles of filaments in thick jellylike sheaths, others fine sheathless filaments, still others elaborately branched filaments that could release spores from their tips.

Finally, many colonial or filamentous cyanobacteria retained nitrogen fixation in their metabolic strategies: they evolved large exceptional cells called heterocysts ("other cells") that were specialized for fixation of atmospheric nitrogen. These cells are not exceptions to the rule that nitrogen fixation is possible only in anaerobic cells. The thickened walls of heterocysts are designed to exclude oxygen, and they do not perform oxygenic photosynthesis—in fact, they rely for food on the molecules that diffuse into them from neighboring photosynthesizing cells.

Freed from their need for hydrogen sulfide and able to use water as a hydrogen donor, cyanobacteria entered many environments—soil, deserts, lakes, streams—and some even penetrated and lived inside bare rocks. Some adapted to cold marine waters, others to hot freshwater springs. Accumulated in ponds and pools, their bodies provided food for saprobes—bacteria that could live on the fixed carbon and nitrogen of dying and dead cells. New intricate food chains developed with cyanobacteria at the bottom as primary producers. The carbohydrate they store like starch, a polymer made up of many interlinked sugar molecules, can be easily broken down to sugars by enzymatic action. Cyanophycean starch, sugars, and even smaller organic compounds became the food for myriads of other cells.

More significantly, the oxygen that cyanobacteria produced enabled other cells to adopt aerobic metabolic modes. Some microbes developed first tolerance, then facultative use, and eventually the obligate use of oxygen in all metabolism. Photosynthetic autotrophs, such as purple nonsulfur bacteria, that had previously evolved cytochromes and electron-transport chains for the anaerobic oxidation of sulfide added the use of oxygen to their metabolic repertoire.

The independent acquisition of oxygen-using metabolism by several microbial groups led to great waves of speciation among prokaryotes. Making use of the greater quantities of energy provided by respiration, prokaryotes diversified into many niches and evolved elaborate specializations and life cycles. Some microbes could bud off motile offspring cells that did not look at all like the parent. These flagellated bacteria would then swim away to a more favorable location, attach to a solid surface, and then "metamorphose" back into the parental form. Some cells formed funguslike networks of living tissue that twined through and around soil particles. Some could form branched structures, called fruiting bodies (by analogy with the much more recently evolved and larger plants), that released spores from their branches. All such elaborate organisms were actually multicellular, although none were eukaryotes.

A Modern Environment

More than a billion years ago, the Earth's geological setting and atmosphere already resembled those of the world we inhabit. Oceans covered some two-thirds of the globe; large continental masses made up the remaining portion of the Earth's surface. Covering the continents were rock outcrops, running water, sand, and soil—all indications of chemical and physical weathering. Mountains thrust upward; glaciation, weathering, transport of sediments, and the formation of lakes were in play.

The diverse and abundant microorganisms made up in chemical activity what they lacked in morphological sophistication. Photosynthesizers, fermenters, chemoautotrophs (which synthesize food by using the energy in inorganic compounds), and diverse heterotrophs interacted with each other and with the life-produced gases of the atmosphere (see Figure 3-8). With the exception of a few fancy compounds, such as the sweet-smelling essential oils of higher plants, the exotic hallucinogens, and the exquisitely effective snake venoms, prokaryotic microbes can produce and break down nearly all the molecules in the repertoire of modern life.

The enormous evolutionary changes that led to animals and plants were yet to come, but the oxygen-releasing cyanobacteria set the stage for their appearance. Animals, for example, could not have evolved before the appearance of photosynthetically produced food and oxygen, upon which they are utterly dependent. However, setting the stage was not enough. Neither

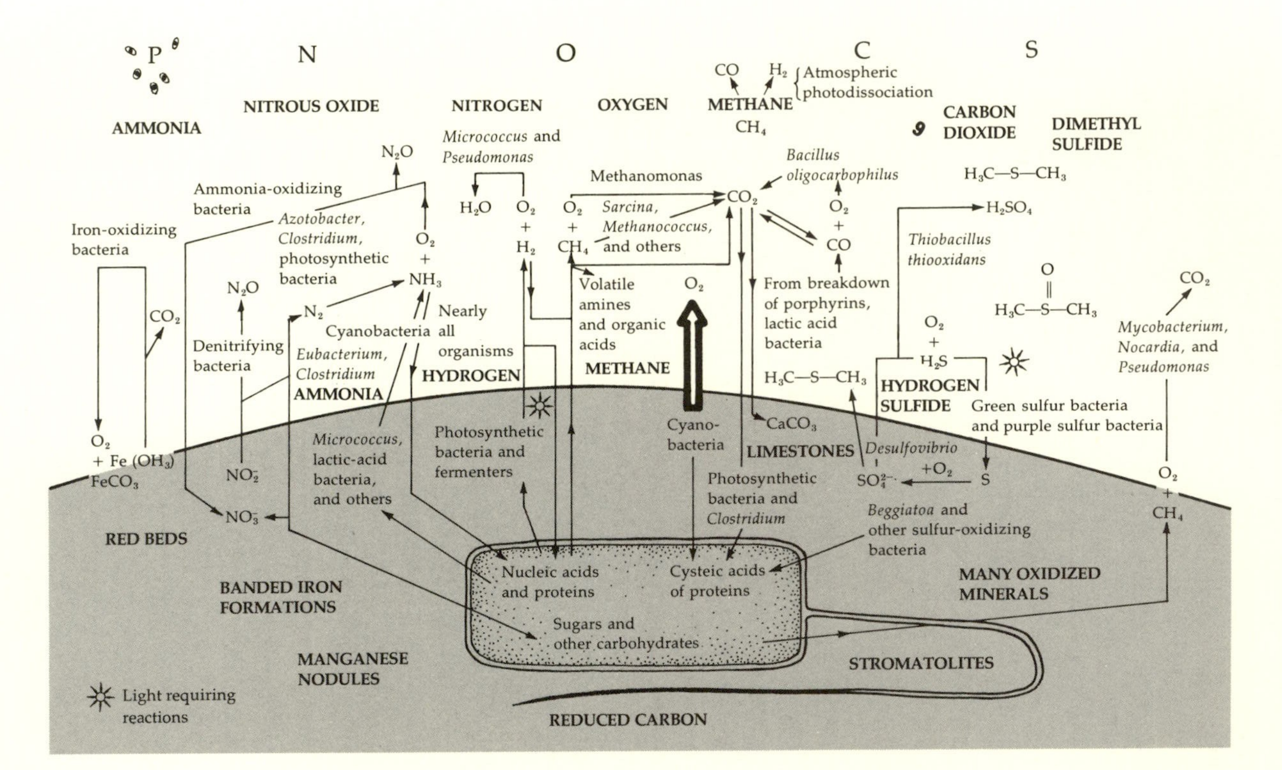

Figure 3-8. A model of the cycling of essential elements in a world having some atmospheric oxygen but inhabited only by bacteria. The dark curved line represents a horizon—atmosphere above, water and sediment below. Under these conditions, phosphorus cannot move through the atmosphere in gaseous form, but it can in the form of phosphates in bacterial spores and other particles. [From L. Margulis, *Symbiosis in Cell Evolution*. San Francisco: W. H. Freeman and Co., 1981.]

animals nor plants could evolve until the type of cell of which they are composed—the eukaryotic cell—came into being. The history of the appearance of this nucleated cell is by no means known in detail, and it may never be. Nevertheless, some strong inferences are possible.

Suggested Reading

Cloud, P. E., Jr. *Cosmos, Earth, and Man: A Short History of the Universe.* New Haven: Yale Univ. Press, 1978.

Fogg, G. E., W. D. P. Stewart, P. Fay, and A. E. Walsky. *The Blue-green Algae.* London and New York: Academic Press, 1973.

Keeton, W. T. *Biological Science,* 3rd ed., Chapter 4, "Energy transformations." New York: Norton, 1980.

Raven, P. H., R. F. Evert, and H. Curtis. *Biology of Plants,* 3rd ed. New York: Worth, 1981.

Stryer, L. *Biochemistry,* 2nd ed. San Francisco: W. H. Freeman and Co., 1981.

Walker, J. C. G. *Earth History.* Portola Valley, Calif.: Science Books International, forthcoming.

CHAPTER 4

A New Kind of Cell

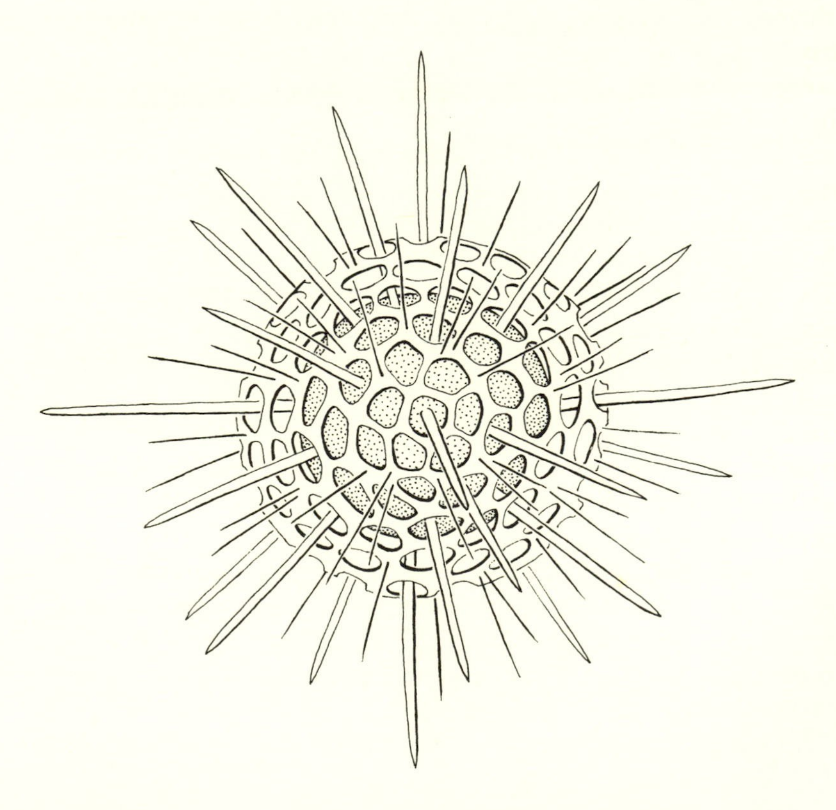

TWO BILLION years ago, the land was covered not by forests, marshes, grasslands, tundra, and chaparral, but by the blue-green and purple scum of photosynthetic bacteria underlain by yellow, brown, and black layers of nonphotosynthetic and anaerobic bacteria. Bacterial fruiting bodies and nets of fungus-like bacteria grew between the soil particles. In the sea only bacteria floated and swam; in the air only spores were blown by the winds.

Large organisms—the kind usually studied in biology rather than in microbiology classes—are all eukaryotic. As a rule, their cells are larger than the nonnucleated cells of bacteria, or prokaryotes; they contain not only nuclei, but also complex organelles and elaborate membrane systems—mitochondria, plastids (in photosynthesizers), undulipodia, Golgi bodies (see Figure 1-1). Two classes of organelle, mitochondria and photosynthetic plastids, contain their own DNA—their own genes. These are even known to reproduce by dividing like independent cells (see Figure 4-1).

Eukaryotic cells are not only structurally more elaborate than prokaryotic ones, they can also carry out more sophisticated survival strategies. Unlike prokaryotes, most of them are capable of phagocytosis ("cell eating," the engulfing of large, solid particles such as whole bacteria) and pinocytosis ("cell drinking," engulfing of droplets of protein). In times of scarcity, some of them can make thick-walled, desiccation-resistant cysts in which they reproduce and wait for better times. Some single-celled eukaryotes form poisoned darts (trichocysts or toxicysts) with which they stab their prey. Unlike prokaryotes, which usually

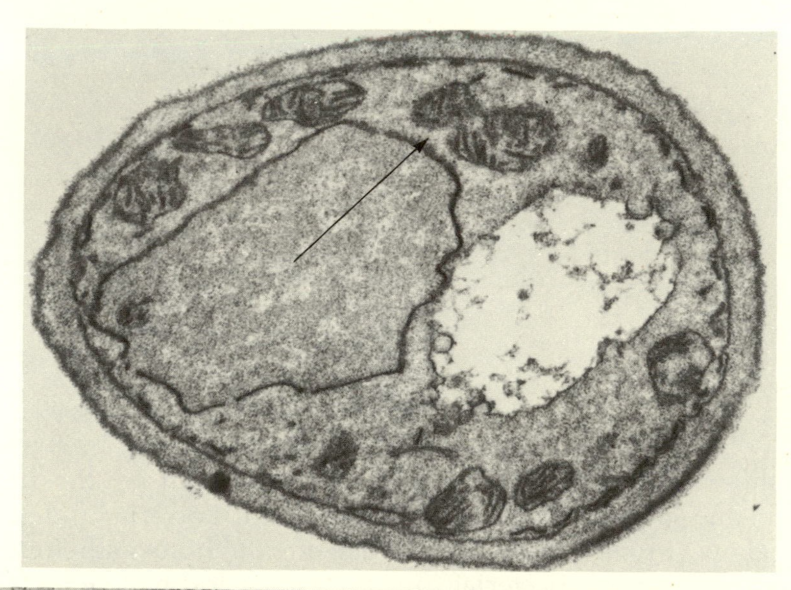

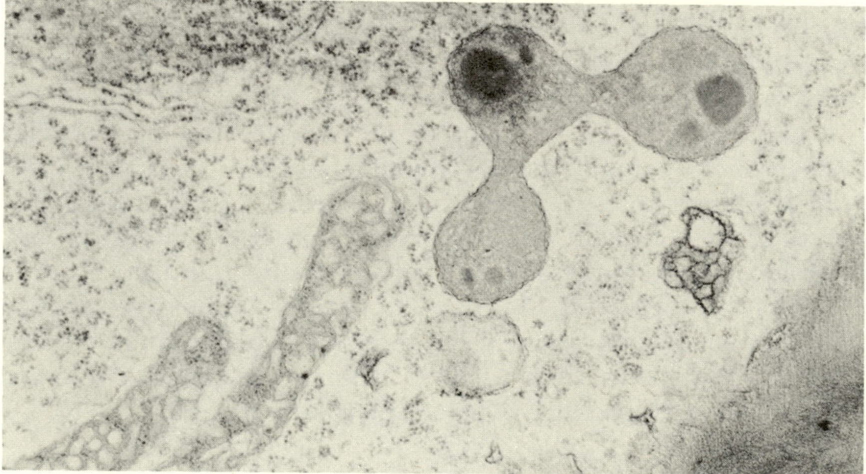

Figure 4-1. (Top) A yeast cell containing several mitochondria (small dark bodies). The one at upper right (arrow) is apparently dividing in two. The cell is about 6 micrometers (millionths of a meter) long. [Courtesy of A. W. Linnane, Monash University, Clayton, Victoria, Australia.]
(Bottom) Part of the fluid-conducting tissue of *Zea mays* (corn). The lobed form at upper right is a chloroplast apparently dividing into three parts. Each lobe is about 0.5 micrometers in diameter. In the two mitochondria at lower left, the convoluted internal membranes are clearly visible. [Courtesy of Michael A. Walsh, Utah State University, Logan.]

contain only a single loop of DNA and reproduce by direct cell division, eukaryotic cells carry their genes on chromosomes and reproduce by the intricate process of mitosis (see Chapter 5). Cell reproduction by mitosis was a prerequisite to the origin of regular two-parent sex.

The First Eukaryotes

What did the first eukaryotic cells look like? Are there organisms today that can be thought of as their direct, unmodified descendants? When did the first eukaryotes appear? None of these questions can be answered with certainty. The first eukaryotes were probably primitive protists: aerobic, aquatic unicellular organisms whose patterns of cell division and life cycle can vary extremely from one species to another. Protists traditionally were considered miniature plants (if they photosynthesized) or animals (if they didn't). Today, with their multicellular relatives, such as marine algae (seaweeds), they are considered to be a separate kingdom, the protoctists, consisting of some thirty major groups. The best known are amoebas, ciliates, green algae, diatoms, dinoflagellates, and red, brown, and yellow-green algae.

Unfortunately, it is nearly impossible to trace in the fossil record the story of the origin and early evolution of protists. Most primitive protists, such as giant amoebas, live in the sea or in pond water. They never form hard parts and their soft bodies break open when they die. All of these properties make the likelihood of fossilization extremely low. Most of the evidence concerning the characteristics and life styles of early protists, therefore, comes from studies (primarily with the electron microscope) of modern ones. Several features of this group point to the protists' position as the first eukaryotes. For example, their idiosyncratic patterns of reproduction suggest that both mitosis and regular two-parent sex evolved within the group.

Lack of convincing fossil evidence makes it difficult to assign a date to the appearance of the first eukaryotic cells. Because protists of various kinds must have evolved before larger and more advanced eukaryotes, they must have done so more than 700 million years ago, when bona fide animals and large seaweeds were living along the shores of ancient seas.

Several theories of the origin of eukaryotic cells have been put forward. Some biologists believe that the various organelles differentiated, or "pinched off," from the nucleus. According to this view, cells first evolved internal membranes, some of which packaged the cell's DNA and RNA in a nucleus. Some of these nuclear genes then escaped from the nucleus and became enveloped by other intracellular membranes. With time, these membrane-bounded genes formed the organelles—mitochondria, plastids, and undulipodia—of eukaryotic cells. Such a scenario does account for the origin of various organelles, but it does not explain another observation: the DNA, the genetic material, of organelles is different from the DNA in the cell nucleus. It is not organized with proteins into chromosomes as the nuclear DNA is—in fact, it is much more like bacterial DNA than like nuclear DNA.

Other biologists, including myself, hold an entirely different view of the origin of eukaryotic cells. We believe that independent, free-living microbes joined together, first casually, then in more stable associations. As time passed, as evolutionary pressures favored such symbiotic unions, the partner microbes became permanently joined in a new cell composed of interdependent components. According to this theory, three classes of organelles—mitochondria, plastids, and undulipodia—once lived as independent prokaryotes, an idea that would account for their separate genetic material.

Biological Partnerships

The cell symbiosis theory is supported by increased understanding of biological partnerships. The importance of such partnerships has frequently been overlooked. Evolutionists in the nineteenth century emphasized the competitive theme of Darwin's work: in a vicious fight for survival, successful organisms leave more offspring, and the strongest competitor is thereby selected. True, the organism ahead in the evolutionary game is the one that leaves the most offspring. Yet there is no single winner of the life game. No individual, in fact no species, can survive in the total absence of all others. Gases and food must be provided, and carcasses and refuse must be carted away, digested, and recycled. These processes require many kinds of organisms.

Symbioses are common among microbes, especially in oxygen-poor environments. In the absence of oxygen, organic matter breaks down slowly. The transformation of a dead bacterium into its final gaseous end products, such as ammonia, hydrogen sulfide, and methane, often requires the activity of more than one species. For example, for years the bacterium *Methanobacillus omelianski* has been cultured from soil as a microbe that makes methane from ethanol. Now the bacterium is known to be two organisms, which look so similar that biologists had not been able to distinguish them until recently. One ferments ethanol to hydrogen and carbon dioxide; the other converts these gases to methane. The ethanol-fermenting prokaryote is primitive, an example of what the earliest cells may have looked like, whereas the second, methanogenic prokaryote is a rather sophisticated natural gas maker. Together, the two organisms grow much faster than they do apart, although skillful microbiologists have been able to grow them separately.

Organisms of all sizes and from all five kingdoms enter into symbiotic relationships. Bacteria live in the guts of animals, fungi live symbiotically with green algae or cyanobacteria in lichens, and nearly every species of echinoderm (starfish, sea urchin, and the like) has one or two of its own protoctist symbionts.

Every green plant also has its associated microbiota: on leaf surfaces, intertwined among root hairs, on twigs, and branches. The small feeding roots of nearly all forest trees are ensheathed by fungi, which function as highly efficient extensions of the root system, transporting essential plant nutrients, such as phosphorus and nitrogen, to the tree roots. And the leguminous plants, a large family that includes peas and beans, have been enormously successful at least in part because of the symbiotic nitrogen-fixing bacteria that inhabit their roots.

The Nucleus

In all eukaryotic cells, most of the genes (hereditary information coded in strands of DNA), are inside the nucleus. Surrounded by protein and RNA, the DNA of the nucleus is packaged in at least two and sometimes in more than a thousand chromosomes, depending on the species. A distinct single-layered membrane encloses the nucleus and separates it from the cytoplasm, the fluid that fills the rest of the cell.

No one is certain what evolutionary selection pressures led to the packaging of genes into nuclei, but three explanations have been suggested. For simplicity, I shall refer to these explanations as the "untangled DNA" hypothesis, the "oxygen protection" hypothesis, and the "symbiotic origin" hypothesis.

DNA replicates by complementary polymerization of nucleotides, as was described in Chapter 1. Dividing eukaryotic cells, which not only are larger than prokaryotic cells, but also carry more genetic material, must keep their very long molecules of newly synthesized DNA untangled, thereby ensuring each offspring cell a complete copy of the DNA. Proponents of the untangled-DNA hypothesis suggest that the cell nucleus first arose as an enveloping membrane in the center of the cell. As replication proceeded, the two resulting DNA molecules could attach themselves to different spots on the membrane, thus keeping separate from each other.

The hypothesis is suggested by what sometimes occurs during cell division in prokaryotes. When prokaryotic cells divide, DNA is partitioned to offspring by attachment to growing membranes, either the outer cell membrane or, in some cases, an elaborate infolding of it called the mesosome (see Figure 4-2). One can imagine, then, that in some ancestral bacteria the mesosome finally broke away permanently from the outer membrane and surrounded the DNA. Once the cell had an internal membrane—now a nuclear membrane—even more DNA could be carried, replicated, and distributed precisely to daughter cells without becoming entangled.

According to the oxygen-protection hypothesis, a membrane-bounded nucleus arose to keep the oxygen required for respiration away from the sensitive genetic material. Eukaryotic membranes, like prokaryotic, are composed of proteins and lipids, but eukaryotic membranes contain several kinds of lipids not found in most prokaryotes. These include the fused-ring compounds called steroids (see Figure 4-3). Steroids require oxygen for their production. By using up small quantities

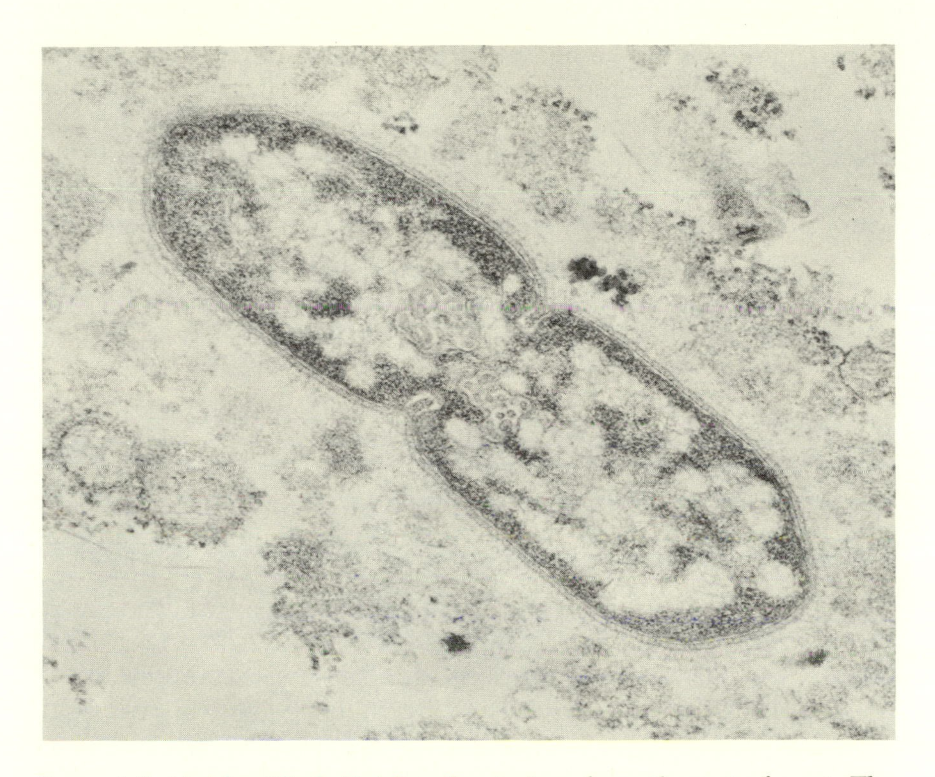

Figure 4-2. Unidentified dividing bacterium from the gut of a rat. The mesosome has already divided; the two daughter mesosomes are on opposite sides of the wall that is forming to divide the bacterium in half. The cell is about 1 micron long. [Courtesy of David Chase, Sepulveda Veterans Administration Hospital, Sepulveda, California.]

CH_3 CH_3 CH_3 CH_3 CH_3 HO

Figure 4-3. Cholesterol, a typical steroid. Although it has a bad reputation because its clogs the arteries of patients who have atherosclerosis, cholesterol is essential to the synthesis of vitamin D and the sex hormones.

of this gas to produce such lipids, eukaryotic cells could keep oxygen away from their genetic material. The need for such segregation would explain why steroids were selected for initially. Later, they served cells in other ways as well. For example, because eukaryotic membranes contain steroids, which "lubricate" the membrane proteins, they are flexible, they break and fuse readily, form vesicles, and wrap organelles.

J. Goksør, a Norwegian botanist, J. Pickett-Heaps, a plant-cell biologist at the University of Colorado, and H. Hartman of the Massachusetts Institute of Technology have reconsidered a hypothesis mentioned at the beginning of this century by the Russian biologist K. C. Mereschkowsky. This is the idea that the nucleus originated symbiotically—it is the remnant of a symbiont that invaded a host cell and then lost most of its cytoplasm. This theory needs more detail. What kind of symbiont entered what kind of host? What happened to the host cell's DNA?

Proponents of the theory must also explain the joint functioning of the nuclear and cytoplasmic genetic machinery. In all eukaryotes, DNA in the nucleus directs—by the mediation of messenger RNA—the synthesis of proteins by ribosomes floating in the cytoplasm. I believe that the nucleus and cytoplasm are part of a coordinated, integrated genetic system and that they always have been. Therefore, given the present evidence, I do not see the need to involve symbiosis in the origin of the nucleus, although I believe it played an important role in the evolution of other organelles.

The first two hypotheses—that nuclear membranes differentiated inside cells to enable newly synthesized DNA to be distributed properly to offspring cells and to protect the genetic material from oxygen—seem to me to be correct, and I suspect that both requirements played roles in the evolution of the nuclear membrane.

However it arose, the nuclear membrane probably had a direct influence on the development of chromosomes, which are densely coiled complexes of DNA and protein. The phosphate groups of DNA strands tend to be ionized (negatively) and so repel one another. Thus, DNA would be expected to resist close packing. However, the fluid of a cell nucleus has a concentration of sodium ions considerably higher than the rest of the cell, and the sodium ions (positive) tend to neutralize the mutually-repelling phosphate groups. Thus, the special environment enclosed by the nuclear membrane may have favored the formation of chromosomes. In most eukaryotes now the negative charge of the DNA molecules is neutralized by the positive charge of the histone chromosomal proteins.

Mitochondria

Mitochondria are small, distinct, membrane-bounded bodies —typically ovoid—that float in the cytoplasm (see Figure 4-4).

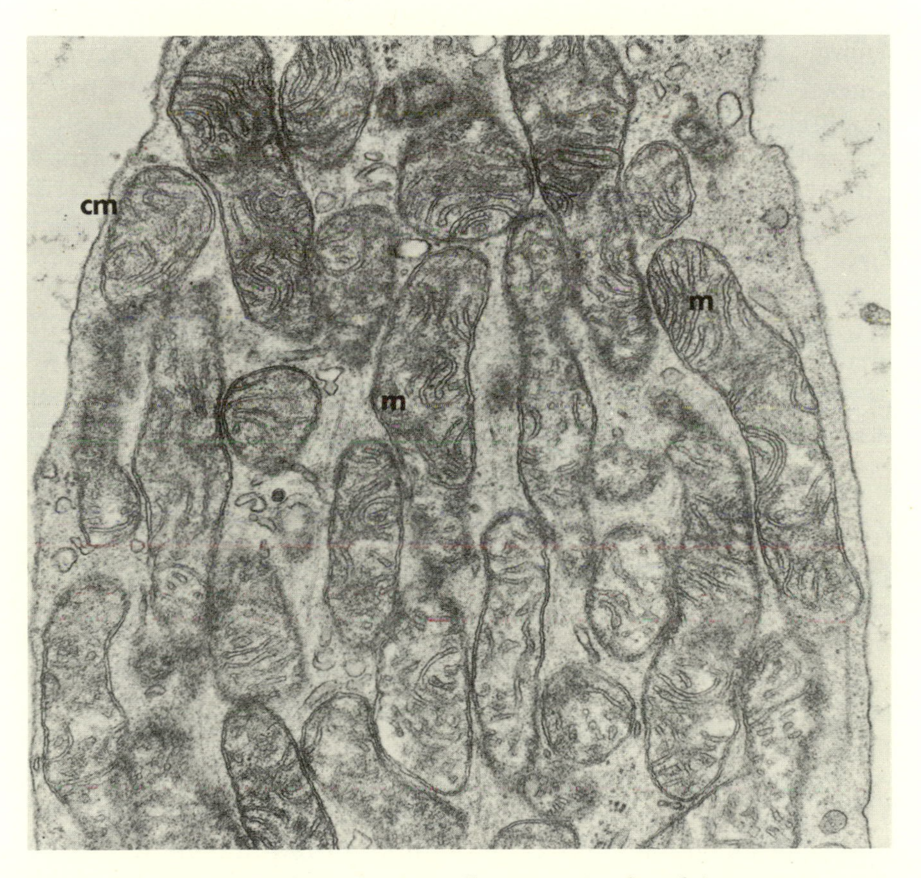

Figure 4-4. Mitochondria in a rat cell. m = mitochondrion; cm = outer membrane of the cell. Each mitochondrion is about 1 micrometer (millionth of a meter) long. [Courtesy of David Chase, Sepulveda Veterans Administration Hospital, Sepulveda, California.]

Typically, eukaryotic cells contain many mitochondria (although certain cells do have only one). All the respiratory enzymes of eukaryotes can be found in the mitochondria, and they are potent generators of ATP. The evolution of mitochondria, therefore, increased the efficiency of energy generation within cells.

Unlike prokaryotes, which can ferment many different organic compounds or even extract energy from certain inorganic compounds, respiring eukaryotes all derive energy in a remarkably similar way. Typically, metabolism begins in the cytoplasm, where food molecules such as glucose are fermented to pyruvic acid. The pyruvic acid molecules are then tranformed into a compound called acetyl coenzyme A, which enters the mitochondria. There, within the mitochondria, the cyclical series of reactions known as the citric-acid or Krebs cycle takes place. Some of these reactions remove carbon from organic compounds and release it as carbon dioxide. Consequently, eukaryotes release carbon dioxide as a waste product of respiration.

During the Krebs cycle, hydrogen atoms are removed from small carbon compounds and transferred to an electron-transport chain composed of the same kinds of compounds, cytochromes and others, that make up the electron-transport chain in aerobic bacteria. Eukaryotic electron-transport chains are remarkably uniform. No matter what the organism, the final electron acceptor is oxygen, which, upon receipt of electrons, becomes reduced to water.

Eukaryotic cells use ATP just as prokaryotes do, for driving the reactions that build up cell material. ATP is the cell's main energy carrier: some of the early steps in glucose breakdown, for example, require ATP to "prime the pump." In addition, the many types of eukaryotic cell movement—intracellular streaming, the beating of undulipodia (cilia and eukaryotic flagella), the contraction of muscle, and the movement of chromosomes in mitosis—all require ATP. When ATP is depleted, these movements stop—they "run out of gas" until more ATP is generated. The requirement for ATP to power movement directly is characteristic of eukaryotes; prokaryotic motility requires neither oxygen nor ATP.

The enzymes catalyzing the respiratory, ATP-yielding transformations are arrayed on the inner layers of the mitochondrial membranes (see Figure 4-5). Because most of the energy of the original glucose molecule is extracted and most of the ATP is

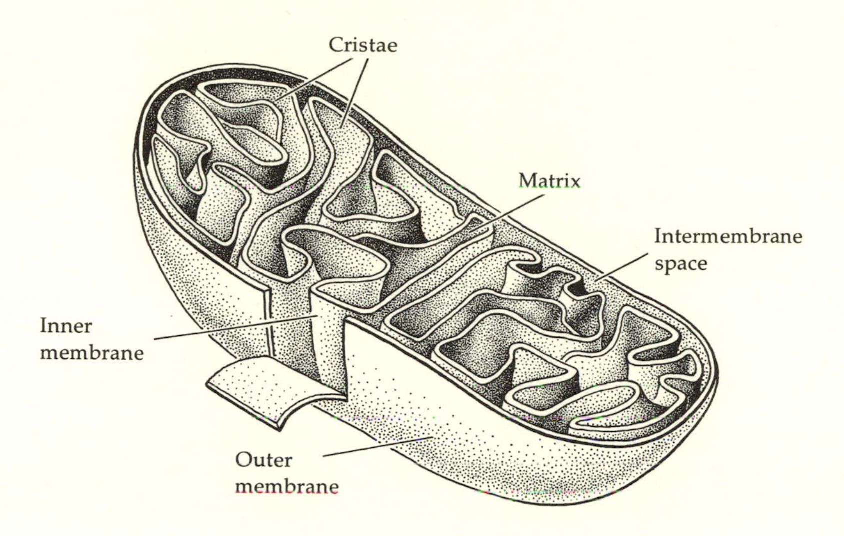

Figure 4-5. A cutaway view of a typical mitochondrion. The outer membrane is continuous with the endoplasmic reticulum and the membrane of the cell nucleus (see Figure 1-1). The cristae (the folds of the inner membrane) bear the molecules of the respiratory electron-transport chain. The reactions of the Krebs cycle take place in the matrix, the space surrounded by the inner membrane. [Drawing by Laszlo Meszoly.]

generated on the inside of the mitochondria, these organelles have been dubbed the "powerhouses of the cell." However, mitochondria are more than powerhouses: they also contain some of the enzymes required for the synthesis of steroids. The first steps in the steroid synthetic pathway take place in the cytoplasm, but the last ones take place on mitochondrial membranes; the two systems cooperate in the synthesis of the final product.

How did mitochondria evolve in early eukaryotic cells? Eventually, as atmospheric oxygen increased, nonphotosynthetic, respiring microbes arose that were entirely dependent on oxygen for metabolism. I believe that such free-living aerobic bacteria containing respiratory systems similar to those of mitochondria established first casual, then more stable, and eventually permanent liaisons with larger anaerobic bacteria. Perhaps the first casual relationship between the aerobes and anaerobes was that of predator and prey (see Figure 4-6). Some of the anaerobes would have evolved tolerance to their predators, who would then come to reside for extended periods in the

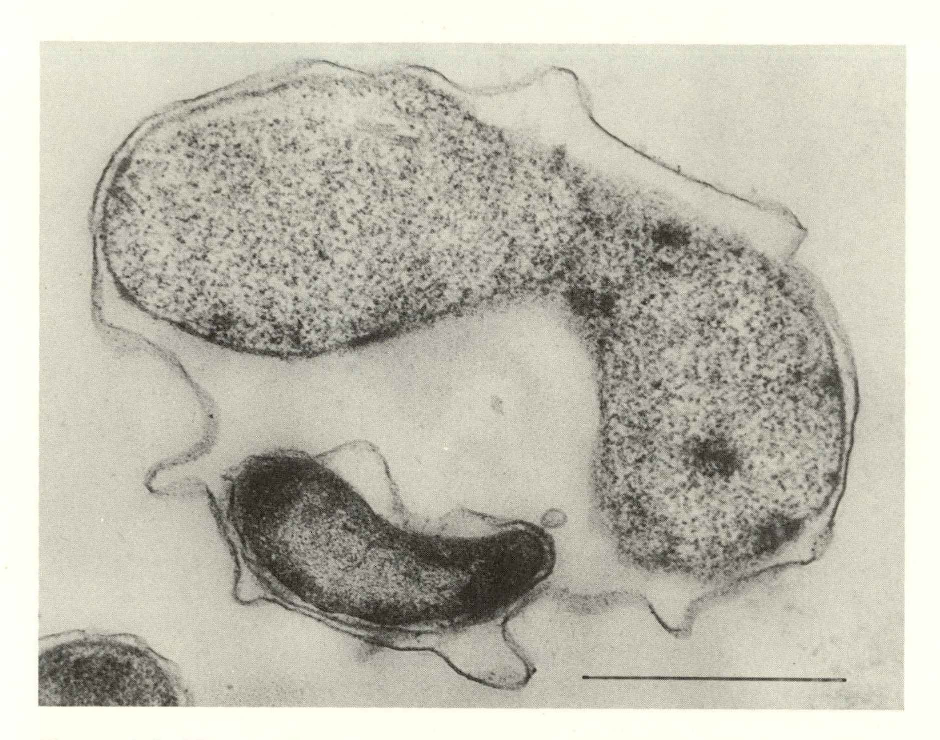

Figure 4-6. The predatory bacterium *Bdellovibrio* (smaller and darker body) inside the cell of another species of bacterium. The *Bdellovibrio* lives in the space between the host's cell wall and the flexible membrane on its inner side. The bar represents a length of 0.5 micrometer (millionth of a meter). [Courtesy of H. Stolp, Bayreuth University.]

food-rich interior of the host cells. Tolerance would have required, at the least, protecting the host's DNA from the oxygen that the guest aerobes needed to live and grow. Thus, if the oxygen-protection hypothesis is correct, the host cell would have evolved a nuclear membrane to shield its DNA from its poisonous guests.

As the concentration of oxygen in the atmosphere increased, such partnerships would have been favored. Unlike its anaerobic relatives that had not developed nuclear membranes, the host cell was not restricted to the shrinking number of anaerobic living environments; further, it could use the energetic products of its partners' efficient aerobic metabolism. The small aerobic partners, meanwhile, lived in a rich soup—the waste products of the host's fermentation—and were protected from the dangers of a free life. Eventually, the hosts came to depend on their former enemies and the guests entirely gave up their independent life, becoming the first mitochondria. Mitochondria thus represent cells inside cells. As symbiont and host cell became progressively more dependent on each other, they evolved into the modern eukaryotic cell.

The symbiotic theory of the origin of mitochondria is not new. Cytologists first described mitochondria in the 1840s, following the introduction of the high-power compound light microscope. Even when viewed under instruments quite crude by today's standards, mitochondria looked so much like bacteria that some early cytologists proposed that mitochondria once had been respiring, free-living microbes. The foremost advocates of symbiotic theories were the Russian K. C. Mereschkowsky (1909), the Frenchman P. Portier (1919), and the American I. Wallin (1927).

The symbiotic theory rests on a firmer basis today. Work at numerous laboratories has shown that mitochondria from all organisms contain the essential components of bacterial replicating systems. Mitochondria have their own DNA, messenger RNA, transfer RNA, and ribosomes. All these components are enclosed by mitochondrial membranes.

Another line of evidence suggesting separate evolutionary lineage comes from the study of cell sensitivity to antibiotics. Mitochondrial ribosomes and those of respiring bacteria are sensitive to the same antibiotics (see Table 4-1).

Once it became a symbiont, the mitochondrion began to lose its genetic independence: the amount of DNA and RNA began to decline. Some of the decrease probably resulted from natural selection against redundancy, which often accompanies the evolution of symbioses. For example, if host and symbiont organisms both synthesize a required nutrient, with time one of the symbionts will lose its capacity to do so, thereby becoming more and more dependent on the other. Mitochondria in rat liver cells, for example, have less than 10 percent as much DNA as most bacteria.

Today the two genetic systems—nuclear and mitochondrial—are extremely interdependent. For example, in the mold *Neurospora* the nuclear genes code for the enzyme RNA polymerase, which hooks up nucleotides into RNA molecules inside the mitochondria. Although the nuclear DNA specifies the amino acid sequence of the RNA polymerase, which is synthe-

*Table 4-1. Antibiotic Sensitivities of Ribosomes from Different Sources**

	Mitochondria	Bacteria	Eukaryotic cytoplasm
Puromycin	+	+	+
Thiostreptin	+	+	−
Cycloheximide	−	−	+
Anisomycin	−	−	+
Chloramphenicol	+	+	−
Erythromycin	+	+	−
Lincomycin	+	+	−
Neomycin	+	+	−
Streptomycin	+	+	−
Tetracycline	+	+	−

*+, sensitive such that amino acid incorporation into protein is blocked in the presence of small concentrations of this antibiotic. −, not sensitive such that incorporation proceeds at the same rate and to the same extent in the presence or absence of this antibiotic.

sized in the cell cytoplasm, the enzyme is part of the machinery of the mitochondrial genetic system.

Eukaryotes without Mitochondria

One way to learn about functions performed strictly by mitochondria is to study eukaryotes that lack these organelles. The only eukaryotes that do not have mitochondria at any stage in their life cycle are certain obscure protists, some of which live symbiotically in the hindgut of insects. Some of these microbes—hypermastigotes—are highly motile and are not primitive eukaryotes (see Figure 4-7). That is, they did not evolve before the origin of mitochondria; rather, they lost their mitochondria after they began to live under the anaerobic conditions of the insect gut. Some termite-gut symbionts have free-living relatives that do contain mitochondria, and many of both kinds have well-developed mitotic systems, suggesting that they are not primitive relict protists.

If, as it seems, all eukaryotes require the products of mitochondrial metabolism, how do termite-dwelling protists

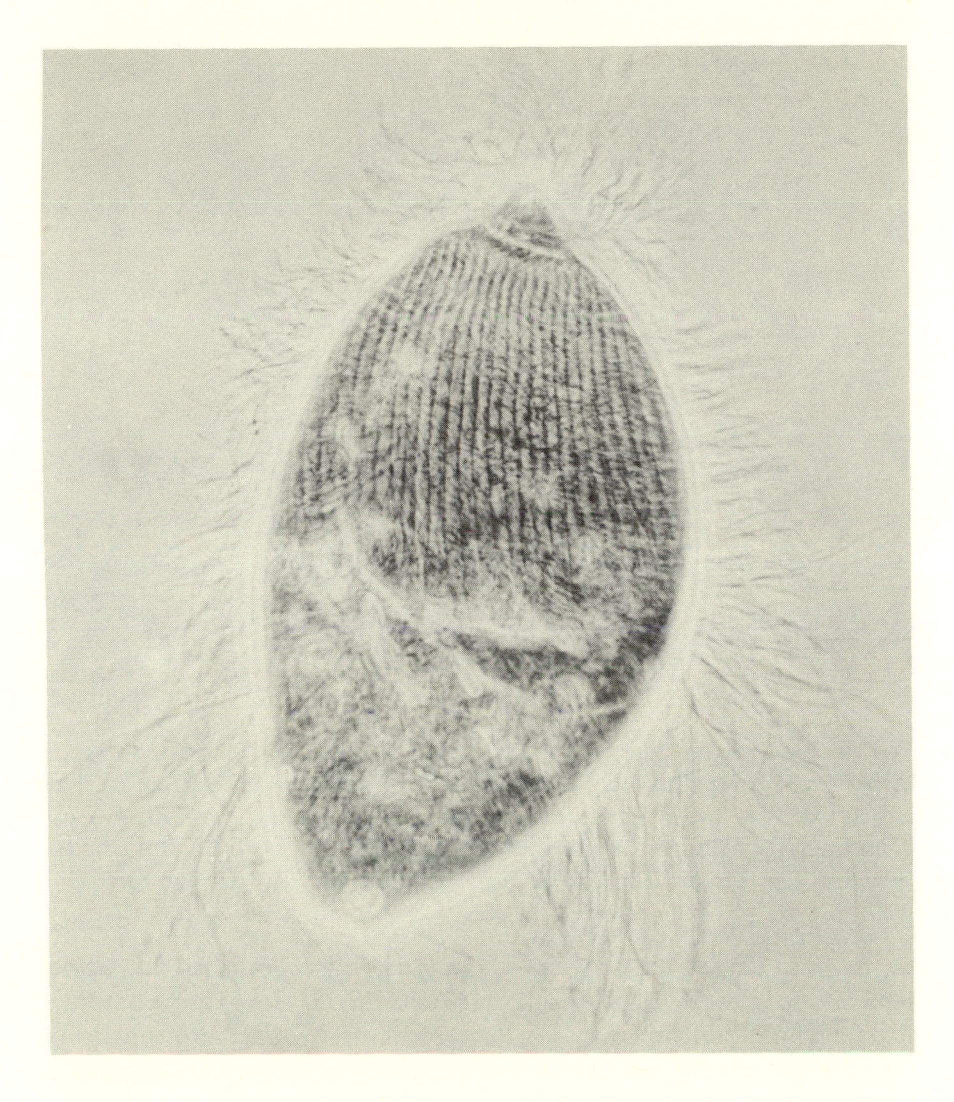

Figure 4-7. Trichonympha ampla, a protist that lives in the hindgut of the termite *Pterotermes occidentis,* is covered with thousands of undulipodia (eukaryotic flagella). This specimen is about 0.2 millimeter long. [Courtesy of David Chase, Sepulveda Veterans Administration Hospital, Sepulveda, California.]

manage to obtain these products? Apparently, they have mitochondria surrogates; they themselves are hosts to internal symbiotic bacteria that look and act like mitochondria (see Figure 4-8). These symbiotic bacteria, of the same size and distributed

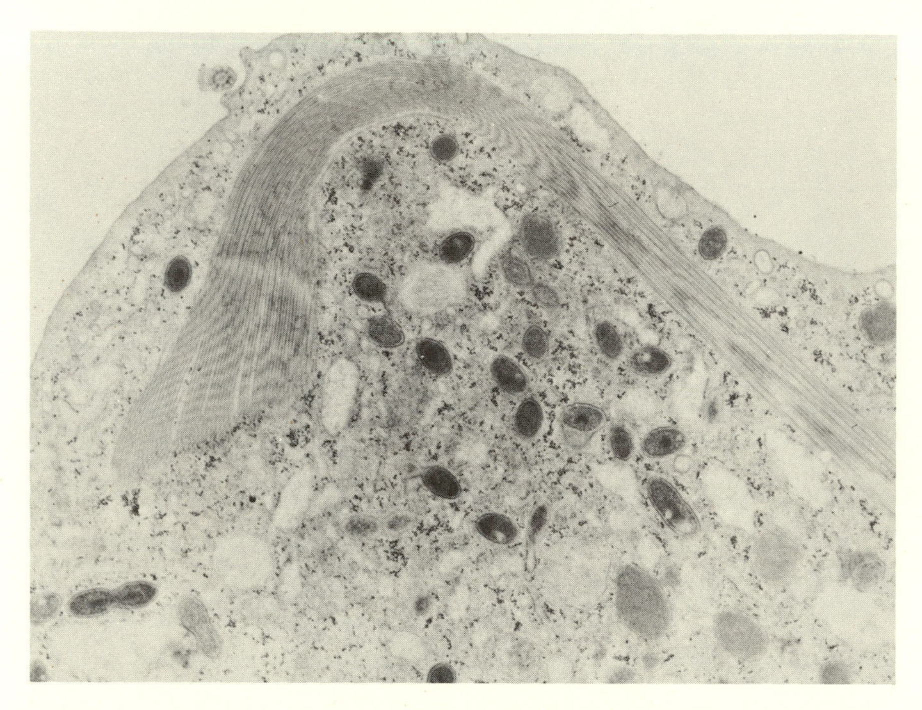

Figure 4-8. Bacteria (dark ovals) acting as mitochondria surrogates in *Pyrsonympha,* a protist that lives in the hindgut of termites. Notice that a bacterium at the lower left is dividing. The bacteria are about 0.25 micrometer (millionth of a meter) long. [Courtesy of David Chase, Sepulveda Veterans Administration Hospital, Sepulveda, California.]

in the same way as mitochondria, probably perform the same service for their hosts that mitochondria do—they oxidize foodstuffs. However, they are better adapted to low ambient concentrations of oxygen (as in the termite gut) than mitochondria would be.

A genuinely primitive eukaryotic organism is *Pelomyxa palustris.* A giant amoeba found in the mud at the bottom of ponds, it also has mitochondria surrogates (see Figure 4-9). Hundreds of individual symbiotic bacteria of at least two types (probably two species) live clustered around the cell's many nuclei and scattered in its cytoplasm. Whether *Pelomyxa* never had mitochondria or it has lost them is a difficult point to resolve because it has no close relatives for comparison. It seems to have solved the problem of aerobic respiration independently of other

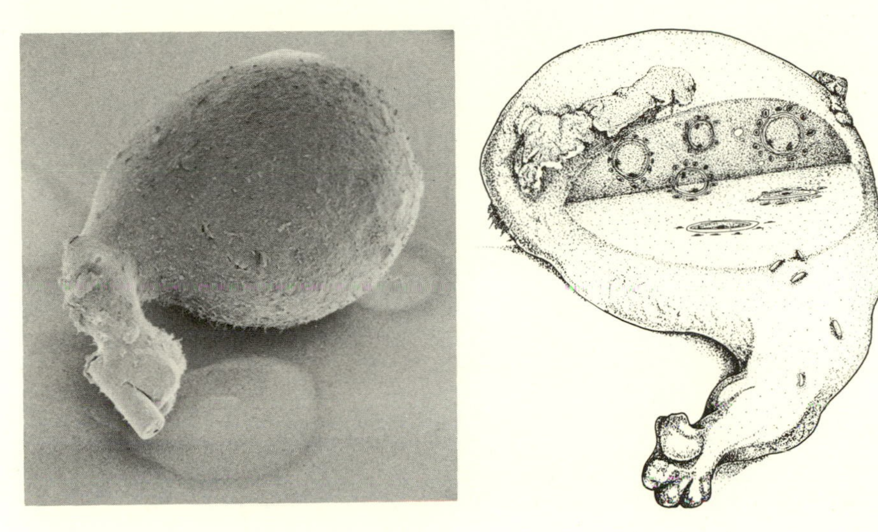

Figure 4-9. (Left) *Pelomyxa palustris*, a giant amoeba that lives in pond mud and lacks mitochondria. This specimen is about 0.25 millimeters long. [Courtesy of E. Daniels, Argonne National Laboratories.]
(Right) In this drawing, the bacteria that serve as mitochondria in *P. palustris* cluster around the several nuclei of their host. [Drawing by Robert Golder.]

eukaryotes. If *Pelomyxa* is treated with antibiotics that kill its endosymbiotic bacteria, lactic acid accumulates in its cytoplasm. This suggests that *Pelomyxa* depends on its mitochondria surrogates to fully oxidize the end products of glucose fermentation, just as mitochondria oxidize them.

Hypermastigotes and other protists that have lost their mitochondria and also lack mitochondria surrogates are unable to form Golgi bodies, highly folded specialized sections of the general eukaryotic internal membrane system (see Figure 4-10). They are composed of the usual membrane steroids and other lipids, but are the repository of other cell products as well. The Golgi apparatus is best developed in cells that secrete specialized proteins such as digestive enzymes and glandular mucus. These products are synthesized on the membranes of the Golgi apparatus; mature parts of the apparatus then separate from it to form vacuoles, closed membranous bags containing the products. Eventually, the vacuoles migrate to the cell's outer membrane and fuse with it, releasing their contents to the outside.

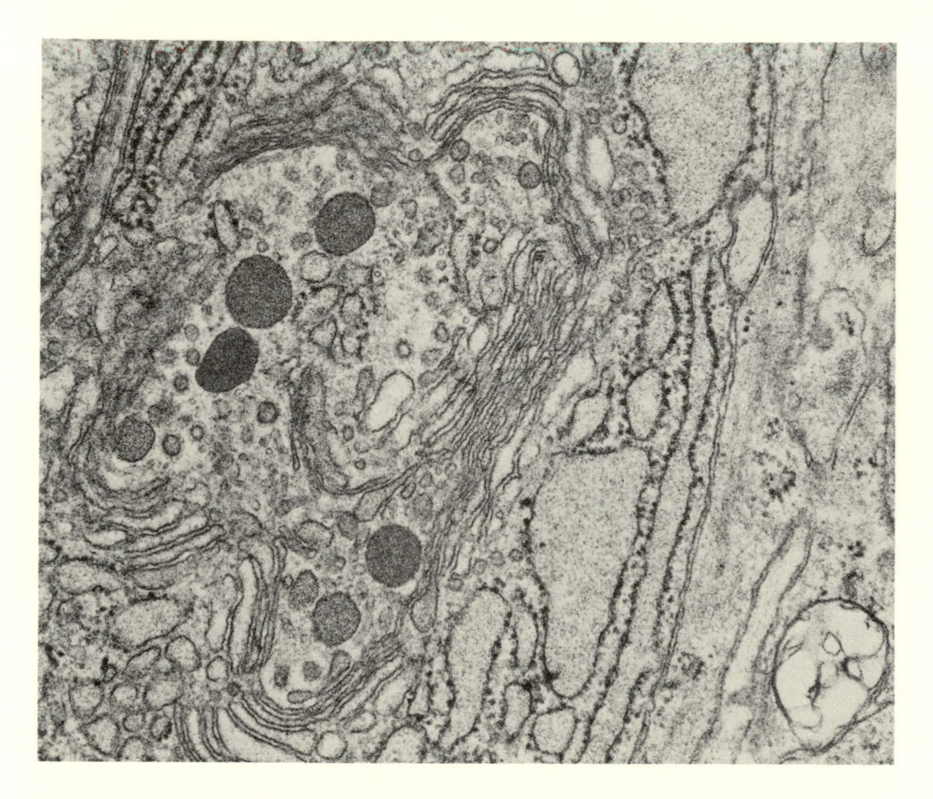

Figure 4-10. The Golgi body in a cell from the intestine of a mouse. [Courtesy of David Chase, Sepulveda Veterans Administration Hospital, Sepulveda, California.]

Organisms or cells not performing such functions contain less-developed Golgi membranes or lack them altogether. All prokaryotes lack a Golgi apparatus, and all eukaryotes that lack mitochondria (or their surrogates) also lack one. The almost perfect correlation between the presence of Golgi apparatus and the presence of mitochondria (or their surrogates) suggests that some mitochondrial products (such as steroids) are required for the formation and functioning of the Golgi apparatus. Thus, mitochondria probably do far more than supply the cell with large quantities of ATP. Although they might have originated as independent symbionts, they are surely an integral and necessary part of the cell now.

Plastids

Once mitochondria were well integrated with their hosts, some of the new eukaryotic cells acquired another internal symbiont, a photosynthetic prokaryote, thus adding photosynthesis to their energy-generating repertoire. Like the bacteria that became mitochondria, the photosynthetic partners eventually became completely dependent on their hosts. Eventually, they evolved into plastids, the pigmented, membrane-bounded organelles found in all photosynthetic eukaryotes (see Figure 4-11).

The first photosynthetic eukaryotes probably were algae. The term *algae*, which translates simply as "water plants," describes a way of life rather than a cohesive and related group of

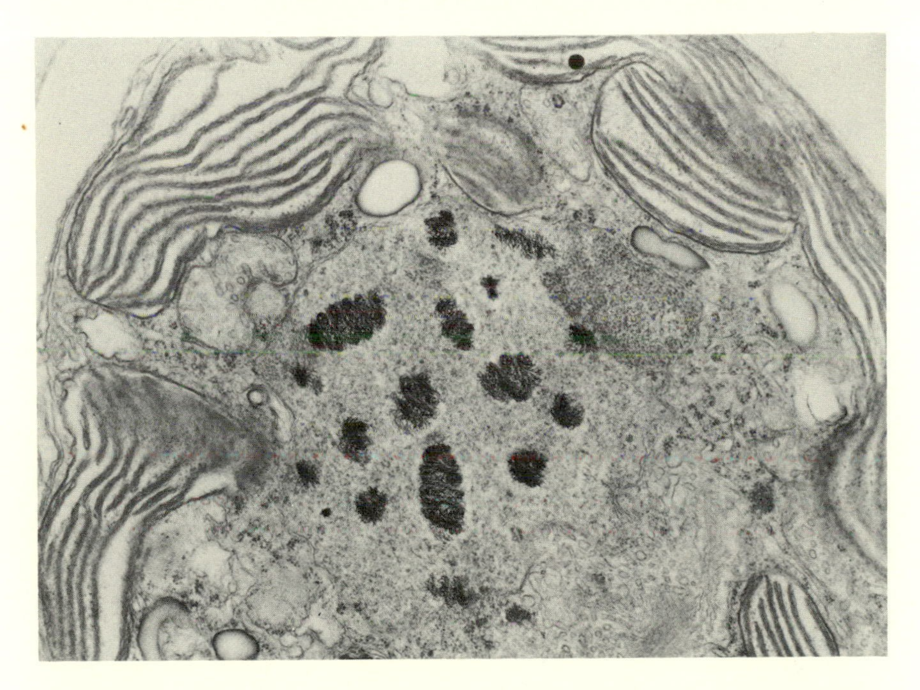

Figure 4-11. Plastids in *Symbiodinium microadriaticum*, a dinoflagellate protist that lives symbiotically in many corals. Notice the stacked layers of thylakoids in the plastids. The dark bodies in the center of the photograph are the chromosomes of *Symbiodinium*. The plastids are about 2 micrometers (millionths of a meter) long. [Courtesy of R. Trench, University of California at Santa Barbara.]

organisms. Traditionally, the term has referred to aquatic organisms that make their living photosynthetically and give off oxygen. Many algae are microscopic single cells; others, such as the giant kelp, are among the largest living organisms. Once classified as members of the plant kingdom, algae are now considered to be protoctists. They can be distinguished from plants in many ways, above all in that they do not form multicellular sexual organs and embryos. Many single-celled algae have protist relatives that are like them in every way except that they lack plastids and therefore cannot produce food photosynthetically. Many algae are highly idiosyncratic. *Caulerpa*, for example, the "turtle grass" that grows in tropical seas, has structures that look deceptively like roots, shoots, and leaves (see Figure 4-12). This organism, which may grow to more than 1 meter in height, is in fact a single huge cell that contains millions of mitochondria,

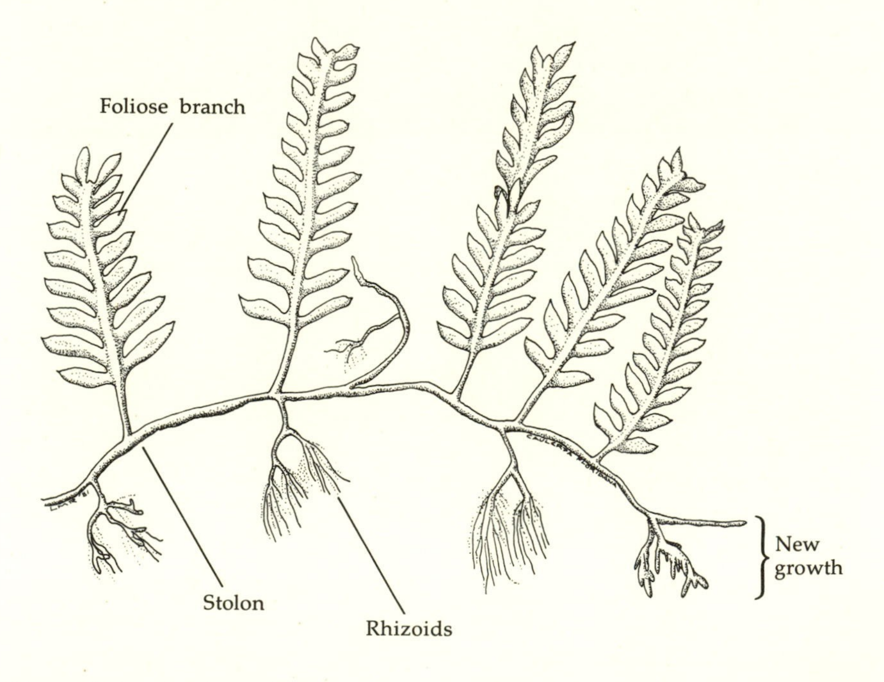

Figure 4-12. Caulerpa floridana, a tropical marine alga, forms many superficially plantlike structures, but is actually a single giant cell. The foliose branches may grow to be more than 50 centimeters long. [Drawing by Linda M. Reeves.]

nuclei, and chloroplasts (green plastids), all bounded by the same single outer membrane. One could hardly consider *Caulerpa* a plant.

The time of origin of the first eukaryotic algae is difficult to determine. By the Ordovician Period, some 500 million years ago, well-developed multicellular algae in the form of seaweeds were thriving underwater. Red algae also appeared early on the scene. Microscopic evidence is more equivocal, but macroscopic organic sheets resembling the seaweed *Ulva* have been found in the rocks of the Dal Group, about a billion years old, in Canada.

The photosynthetic pigments of all algae are bound to membranes inside plastids. Various algal groups differ in the color of their pigments. Chloroplasts are green plastids, rhodoplasts are red, phaeoplasts are brown, and chrysoplasts are golden yellow. In all plants, the plastids active in photosynthesis are chloroplasts. In addition, in plant tissue one finds other types of plastids such as amyloplasts, which are unpigmented and store starch, and chromoplasts, which are pigmented but nonphotosynthetic.

All plant cells contain some sort of plastid. Even the nongreen, nonphotosynthesizing cells of flowering plants (in the roots or seeds, for example) contain plastids, as do those plants, such as parasites, that are unable to photosynthesize. No one is certain what functions all these underdeveloped or nonphotosynthetic plastids serve, but they probably produce lipids and enzymes crucial for the rest of the cell.

The joining of nonphotosynthetic organisms with those able to make their food directly from the sun's energy has occurred frequently. The giant clam *Tridacna,* for instance, can grow four feet in diameter, fed partly by photosynthesizing algae that live between the cells of its mantle. The green hydra supplements its diet of tiny crustaceans by basking in the sun to stimulate photosynthesis by its algal symbiotic partners, which live inside the cells of the hydra. Certain sea slugs acquire chloroplasts by feeding on algae. Inside the slug cells, the foreign chloroplasts function quite independently, photosynthesizing food both for themselves and for the slug. Thus, associations may be on many levels: between autotrophs and multicellular heterotrophs, between single-celled forms, and even between animal cells and chloroplasts.

If plastids were acquired through symbiosis and did not originate within algal cells, one would expect to find that, like mitochondria, they still carry traces of their prokaryotic ancestry. Does this expectation hold up? Several lines of evidence indicate that it does. Except for minor differences in the photosynthetic pigments, the processes of photosynthesis in the various types of algae, in plants, and in cyanobacteria do not differ. Both in cyanobacteria and in plastids, chlorophyll and electron-transport chains are arranged on many-layered membranes, or thylakoids. Plastids have their own DNA and RNA; they can synthesize some of their own membrane proteins. The DNA that makes up the genes of plastids resembles the DNA of prokaryotic cells: it is a circular molecule and it is not coated with the proteins (histones) that coat eukaryotic DNA.

Ford Doolittle, a biochemist at Dalhousie University (Halifax, Nova Scotia), has compared the sequence of bases in the ribosomal RNA from the plastids of the red alga *Porphyridium* (a seaweed) to the sequences in ribosomal RNA from two other sources: the coccoid (spherical) cyanobacterium *Synechococcus* and chloroplasts of the unicellular green protist *Euglena*. The RNA in *Porphyridium* chloroplasts turned out to be more like the RNA from these two sources than like the RNA in its own cytoplasm (see Table 4-2). Doolittle has interpreted his data to indicate that the coccoid cyanobacteria are ancestral to the rhodoplasts of red algae. Similar work suggests that the prokaryote *Prochloron*

Table 4.2. Degrees of Similarity Between the Ribosomal RNA of Porphyridium *Rhodoplasts and Ribosomal RNA from Other Sources*

Source	Similarity (%)
Synechococcus (cyanobacterium)	42
Euglena chloroplast	33
Porphyridium cytoplasm	Less than 15

Source: W. F. Doolittle, Dalhousie University. (Based on RNA nucleotide sequence data, simplified. References and discussion are in L. Margulis, *Symbiosis in Cell Evolution* (1981).)

gave rise to the chloroplasts of green algae.* Unfortunately, no one has yet found yellow or brown pigments in free-living prokaryotes, so the origin of the chrysoplasts of golden-yellow algae and of the phaeoplasts of brown seaweeds remains difficult to infer.

The symbiotic theory of the origin of plastids proposes that photosynthesis was acquired many different times and in a refined and ready form by various kinds of nucleated, heterotrophic cells, or protozoa (the traditional term for nonphotosynthetic protists). After acquiring photosynthetic prokaryotes, they became algae. Motile protozoa became motile algae; nonmotile protozoa became nonmotile algae. The motile algae possessed the efficiency, rapid movement, and developmental capabilities of the protozoa, coupled with the photosynthetic capability of aerobic prokaryotes. They also had two ways of generating ATP: by the light reactions of photosynthesis and by respiration. It is no wonder that they began to dominate the oceans and other wet places of the world. As phytoplankton, they have continued this domination up to the present time, and some green algae eventually gave rise to the green land plants.

Undulipodia

Many biologists agree that evidence favors the symbiotic theory of the origin of mitochondria and plastids. I believe that still a third group of organelles, undulipodia, also became associated with the eukaryotic cell by symbiosis. I should mention, however, that only a few biologists agree with my account of the origin of this third group of organelles. L. S. Dillon, for example, rejects symbiosis as an evolutionary mechanism for any cell organelle (see Suggested Reading, page 106).

There is no basic difference between eukaryotic flagella and cilia. If these hairlike cell projections are long and few, like sperm tails, they are called flagella; if they are short and many, like those that project from surfaces of cells in our tracheas

**Prochloron*, a grass-green prokaryote similar in appearance to a coccoid cyanobacterium, has both kinds of chlorophyll (*a* and *b*) found in the plastids of plants and green algae. Cyanobacteria have only chlorophyll *a*.

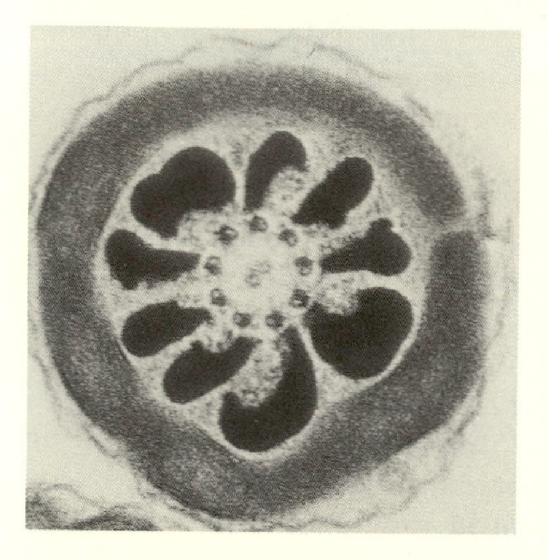
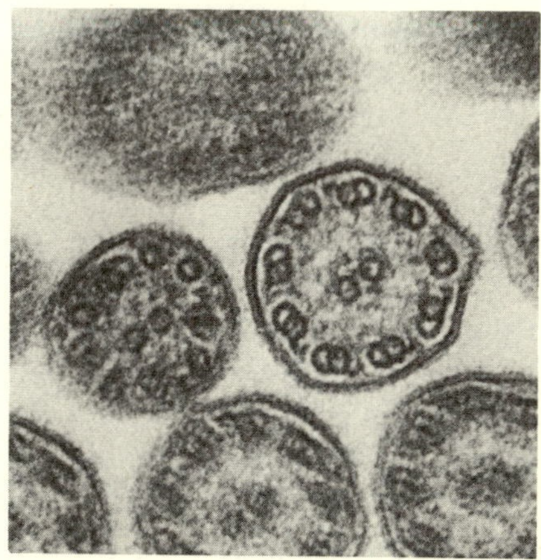

Figure 4-13. (Left) A cross section of the tail of a rat sperm cell. The tail is about one micrometer (one billionth of a meter) wide. [Courtesy of N. J. Alexander, Oregon Regional Primate Research Center, Beaverton, Oregon.]
(Right) Cross sections of cilia on cells that line the oviduct of a rhesus monkey. [Courtesy of Robert M. Brenner, Reproductive Biology, Oregon Regional Primate Center, Beaverton, Oregon.]

(windpipes) and nasal passages, they are called cilia (see Figure 4-13). Their motion propels the cell through its medium or, if the cell is fixed in place, moves particles past it. Tracheal cilia, for example, cause the movement of mucus; their beating is essential for maintaining clean noses and clear throats. With some important exceptions, all protoctists—whether they have plastids and are therefore considered algae or they lack them and are called protozoa—have flagella or cilia. If they are covered with cilia they are called ciliates; if they bear flagella, they have traditionally been called flagellates or mastigotes. Because the term *flagellum* is reserved for bacteria, in this book I will use *mastigotes* for eukaryotic "flagellates." These organelles of motility must have evolved in the earliest eukaryotes. Unfortunately, because early mastigotes had fragile, soft bodies, they left no record of their existence in rocks, so one can only speculate.

Eukaryotic flagella and cillia, whether they come from the sperm of a fern or the nostril of a mouse, are made of bundles of fibers and have a strikingly uniform structure. Seen in cross sec-

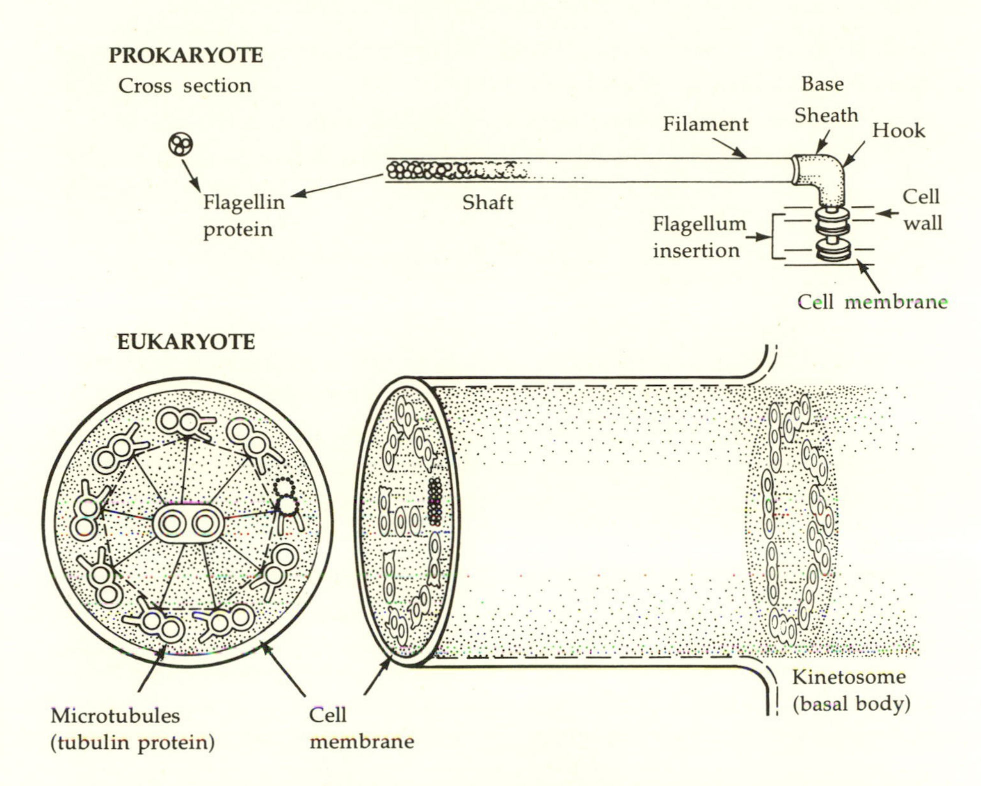

Figure 4-14. The undulipodium (cilium or 9+2 flagellum) of eukaryotes compared with the flagellum of prokaryotes.

tion, all are a quarter of a micron* in diameter and show a circle of pairs of microtubules, minute cylinders (see Figure 4-14). There are generally also two microtubules in the middle of the circle. The entire pattern is known as the 9+2 array. Microtubules from any kind of eukaryotic undulipodium are about 240 angstroms† in diameter and are composed of several similar proteins called microtubule proteins or tubulins. Eukaryotic flagella and cilia invariably grow from structures called basal bodies (the

*One micrometer (also called a micron) = one millionth of a meter. Bacterial cells are usually about one micrometer across.

†One angstrom = one ten-billionth of a meter. A hydrogen atom is about one angstrom in diameter. Ten thousand angstroms = one micrometer.

old term) or kinetosomes, which have nine triplets of microtubules but lack a central pair.

Because eukaryotic flagella and cilia are so similar, it has been suggested that their similarity to each other and their dissimilarity to the far smaller, single-stranded bacterial flagella would be better emphasized if they were called undulipodia ("waving feet"). I shall follow that suggestion here.

Traces of RNA have been found inside the cavity of kinetosomes, a finding that lends support to the view that undulipodia arose through a symbiotic union. It is my hypothesis that undulipodia were once free-living motile bacteria of the spirochete sort (see Figure 4-15). There are many types of these slender, helically undulating bacteria. Some are free living and inhabit watery environments such as mud flats, pond water, and microbial mats, but the better known ones are symbiotic or parasitic, living with insects, mollusks, or mammals. For example, the spirochete *Treponema pallidum* is the organism that causes syphilis in humans.

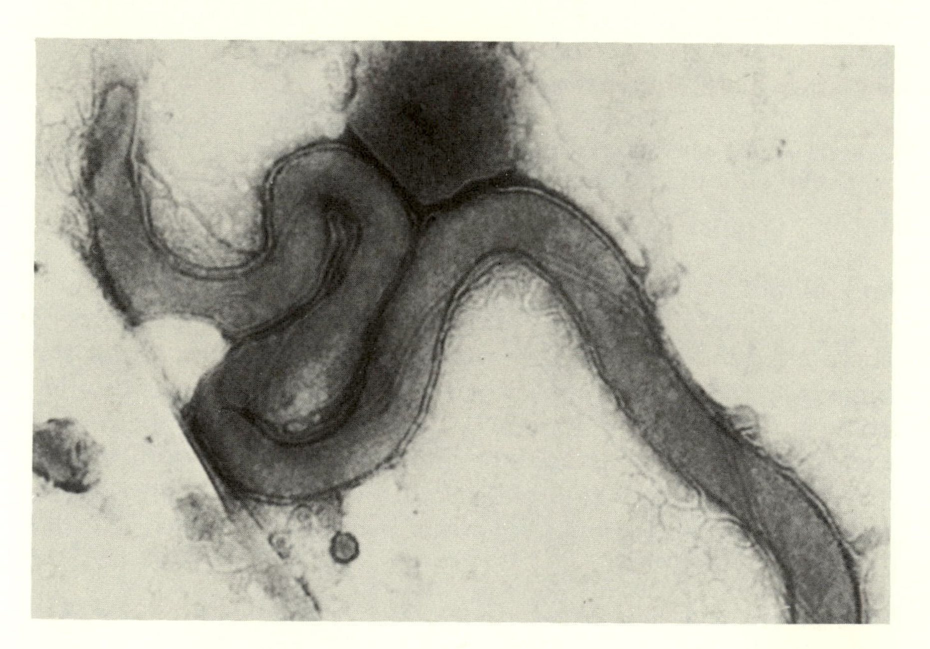

Figure 4-15. A spirochete of the genus *Treponema* taken from the hindgut of a termite. This specimen is about 0.2 micrometer (millionth of a meter) wide. [Electron micrograph by Lelong P. To Isaacs.]

According to my theory, spirochetes formed associations with heterotrophic protists. The spirochetes attached to the surfaces of their protist hosts in order to take advantage of food leaking through the host's outer membrane. Eventually, the undulating spirochetes began to propel their hosts through the aqueous medium. Such relations between present-day spirochetes and hosts, even motility symbioses, are well known. *Mixotricha paradoxa,* for example, a protist that lives in the hindgut of the Australian termite (*Mastotermes darwiniensis*), has four normal 9+2 undulipodia at its forward end. However, these undulipodia take no part in moving the creature. What propels *Mixotricha* are the coordinated undulations of about half a million spirochetes that live on its surface. The four small undulipodia merely serve as rudders, permitting *Mixotricha* to change direction.

Mixotricha is not unique; my colleagues and I have seen at least three other types of motility symbioses among the microbiota that inhabit the hindgut of several different species of termites. Indeed, the spirochetes that live in the hindgut of termites show a natural tendency to attach to objects—living or not—and to beat in unison. In ancient times, then, spirochetes living on a protist's surface may have coordinated to move their host just as their modern analogues do. Thus, they functioned as undulipodia. It is my hypothesis that with the further passage of time, surface spirochetes, some of which by then had become undulipodia, were drawn inside some cells. In certain cells, they never re-emerged. These cells, ancestors to fungi, developed new intracellular processes, including mitosis, that made use of the microtubular components of undulipodia (see Chapter 5).

A similar sequence of events may account for the origin of the amoeboflagellates, a group of organisms that probably resemble the earliest motile eukaryotes. When food is plentiful, these protists withdraw their undulipodia and gorge themselves; after moving around awhile slowly, as amoebas, they divide. When food becomes scarce, however, they develop a typical kinetosome out of which grows an undulipodium; they then become mastigotes, swimming rapidly in search of food. Analogous events are observed in the laboratory. Healthy symbiotic spirochetes can be drawn inside host cells, where they continue to swim around. What they are doing is certainly not clear to observers, unless they are merely showing us that history repeats itself!

If undulipodia do derive from spirochetes, it is easy to see why eukaryotes vary enormously in the number of undulipodia that they bear, but not in the size of each undulipodium. Like all bacteria, the first symbiotic spirochetes had a more or less fixed characteristic size, but their numbers were not fixed. Therefore, one finds as few as a single undulipodium on certain species of water molds and more than a million on some ciliates (see Figure 4-16).

The theory suggests two major corollaries: (1) Researchers ought to be able to find some spirochetes that contain tubulins made up of amino acid sequences similar to those of eukaryotic

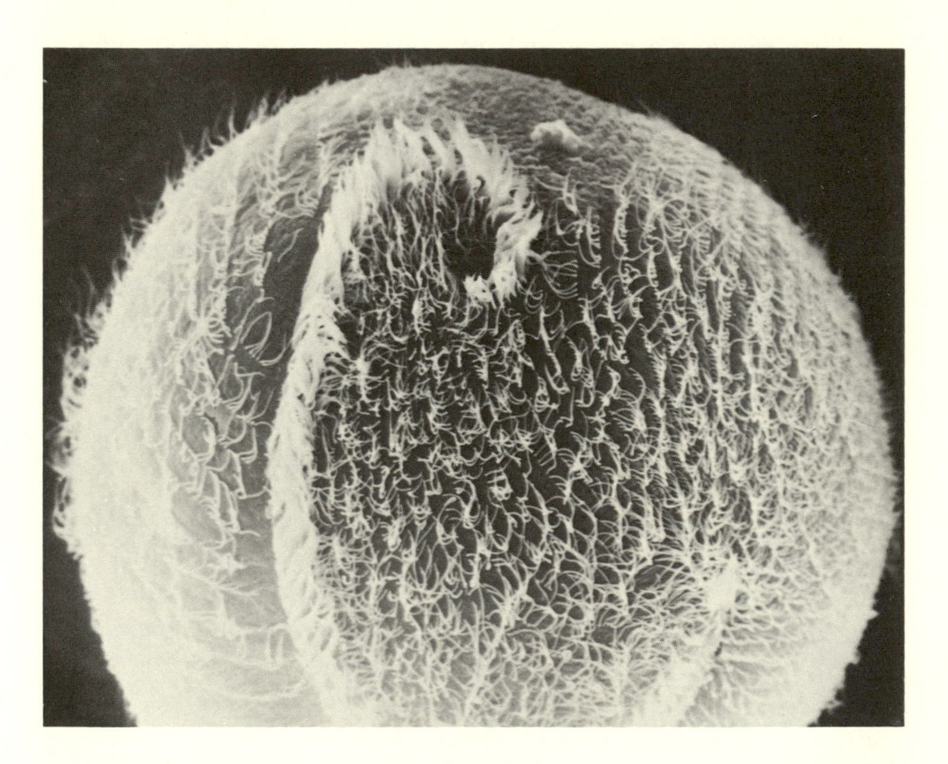

Figure 4-16. The ciliate *Stentor coeruleus,* a common inhabitant of pond water, is covered with thousands of cilia. This specimen is about 0.2 millimeter wide. [Courtesy of Jerome Paulin, University of Georgia.]

tubulins; and (2) if such spirochetes are found, they should contain some DNA and RNA having the same nucleotide sequences as the RNA of eukaryotic kinetosomes. These hypotheses are far from proved, but they are generating interesting experiments. In the mid-1960s, Iran D. Gharagozlou and André Hollande, of the Laboratoire d'Évolution des Êtres Organisés in Paris, discovered that spirochetes inhabiting the hindgut of dry-wood termites contained microtubules 240 angstroms in diameter. In 1977, Leleng To, a graduate student at Boston University, and I, working with David Chase of the Cell Biology Laboratory at the Veterans Administration Hospital in Sepulveda, California, confirmed their observations. We found that the termite spirochete microtubules were smaller, though, about 210 angstroms wide. We also found preliminary evidence that some proteins in these termite-dwelling spirochetes have immunological similarities to brain and undulipodia tubulins. In concordance with the theory, these spirochetes are known to attach to and enter protist cells, and they beat together in a pattern that often looks uncannily like the beating of bona fide undulipodia.

An independent origin helps to account not only for the RNA found inside the cavity of kinetosomes, but also for certain odd characteristics of undulipodia. It would explain, for instance, why undulipodia occasionally swim away from the rest of a protist cell all by themselves. It would also explain why the pattern of undulipodia on certain ciliates can be inherited independently of the nuclear genes.

The theory does have certain weaknesses. Except for the bit of RNA in the kinetosome, undulipodia do not contain their own genetic material. The genes that determine the amino acid sequence in tubulins are on the chromosomes, in the cell nucleus. Thus, the original spirochetes, if such they were, have been drastically modified. They now depend utterly on the living protoctist, animal, or plant cell for their sustenance and for reproduction.

However it happened, the evolution of undulipodia consisting of microtubules was a major advance in the development of cells. As I shall explain in the next chapter, microtubules come into particular prominence during mitosis, the process by which most eukaryotic cells divide.

Movement within Cells

Undulipodia propel the cell itself through its watery environs and move mucus, nutrient solutions, and dust particles along tissue surfaces. Eukaryotic cells also display another kind of movement: intracellular motility. A novice biology student can quickly learn to recognize the difference between large prokaryotes and small eukaryotes by the incessant movement that generally takes place inside the eukaryotes.

In fungi, animal cells, plant cells, and especially in protoctists such as slime molds, the cytoplasm streams around. (Slime molds, often seen as yellow or white jellylike masses on fallen logs, are multicellular or multinucleate protoctists that can grow to be very large.) Organelles, such as mitochondria, and lipid droplets move, often in one-way streams, from the center to the periphery of the cell or mass of slime, and back again. Such intracellular movement, which never takes place in prokaryotes, results from the presence and activity of certain so-called motile proteins, primarily actin, a fibrous protein, and myosin (an ATPase). Neither is a tubulin.

The details of intracellular movement are still under intensive study, but the general mechanism is understood. Filaments made of actin and myosin interact both with each other and with calcium and sodium ions in the surrounding cell solution. When the local concentrations of these ions change, the protein molecules change shape, causing adjacent filaments to slide over each other. The filaments themselves are impermanent—they dissolve and reform continually—and enmeshed particles, organelles, and quantities of cytoplasm are carried along by their movements. Some of these changes require biochemical energy in the form of ATP, as well as ATPases, enzymes that catalyze the breakdown of ATP. Thus, the filaments convert chemical energy into mechanical energy, or movement, and cells control these movements by precisely regulating the local concentration of certain ions—notably, calcium.

Controlled intracellular movement enabled eukaryotes to ingest particulate food, to cycle nutrients within their cells, to expel wastes efficiently, and to regulate the concentration of ions (which entered cells from marine waters). It was also a preadaptation to the formation of muscles and calcium carbonate skeletons. Like intracellular movement, the contraction of muscle is caused by the calcium-regulated sliding of actin and myosin filaments (although the filaments in muscle are more permanent). Thus, intracellular motility and ion regulation must have preceded the rise of the soft-bodied but muscled animals of the Ediacaran fauna some 700 hundred million years ago. Calcium carbonate skeletons, which first appeared in Cambrian times, about 600 million years ago, might have arisen as a by-product of calcium regulation. In fact, an animal that precipitates soluble calcium into a skeleton ensures itself a reserve to be used at a later time. Animals stressed by an unusually high demand for calcium (for example, pregnant women) tend to dissolve their teeth and skeletal calcium rather than draw on the calcium in muscle cells.

The evolution of motility will soon be a solved biological problem. With the reconstruction of evolutionary trees made possible by studies of amino acid sequences in relevant proteins, the entire history of intracellular movement will eventually be unravelled. Actin molecules from rabbit muscle have been sequenced for the first time by a group at the Boston Biomedical Research Foundation, working under the direction of Joseph Elzinga. The place of each amino acid in this actin molecule is known. Brain tubulin molecules have just been sequenced for the first time by Herrick Ponstingl and his team at the Heidelberg Cancer Research Institute in West Germany. Tubulin and actin differ considerably in amino acid sequence. It is only a matter of time until "primitive" tubulin and actin proteins from slime molds, amoebas, mastigotes, ciliates, and other protoctists are purified and sequenced for comparison. Myosin molecules are much larger, and few have been sequenced. There are probably many different types, but most seem to have the ability to break down ATP; that is, they are ATPases.

All the macroscopic advances—the evolution of large algae and animals and the emergence of fungi and plants onto land—required the prior establishment of eukaryotic cells. The evolution of nucleated cells occurred some time between 2 billion and 0.8 billion years ago, but the delicate events themselves have left no direct record. What is certain is that the new cells were larger and more intricate in form than their prokaryotic ancestors. They could also perform more complex functions, including mitotic cell division, the crucial genetic step toward further increase in size. Without mitosis, there could be no meiosis, the kind of cell division that gives rise to sperm and eggs. Without meiosis there could be no animals, plants, or fungi and no obligate association of two-parent sex with reproduction. How the cell acquired sexual sophistication is the subject of the next chapter.

Suggested Reading

Dillon, L. S. *Ultrastructure, Macromolecules, and Evolution*. New York and London: Plenum Press, 1981.

Margulis, L. *Symbiosis in Cell Evolution*. San Francisco: W. H. Freeman and Co., 1981.

Richmond, M. H., and D. C. Smith. *The Cell as a Habitat*. London: Royal Society, 1979.

Whitehouse, H. L. K. *Towards an Understanding of the Mechanism of Heredity*. New York: St. Martin's, 1973.

CHAPTER 5

The Evolution of Sex

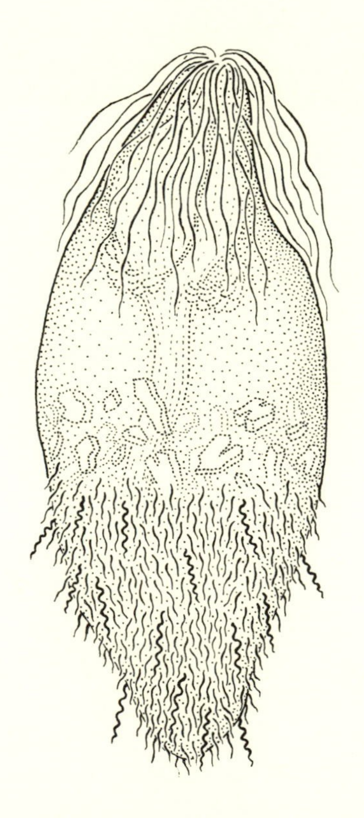

ANY THEORY of the origin of eukaryotic organisms from prokaryotes must account for the major differences between their reproductive systems. All prokaryotes reproduce asexually by binary fission or, less frequently, by budding. A cell divides into two equal offspring cells or an offspring buds off from a parent cell. Modes of asexual reproduction are shown in Figure 5-1. In all these cases, the offspring has only one parent and receives all its genes from that single parent. Prokaryotic sex (see page 49) is never necessary for prokaryotic reproduction. Eukaryotes, however, with the major exception of many protoctists, almost all reproduce sexually; their offspring contain genes from two parents.

The evolutionary path that led from asexual reproduction to two-parent sex was circuitous and full of dead ends. It included steps such as the organization of genetic material into chromosomes and the development of mitosis, the mode of cell division that ensures the precise replication and distribution of chromosomes to daughter cells. The appearance of mitosis made possible the development of a variant—meiosis. Like mitosis, meiosis begins when the chromosomes replicate themselves. In meiosis the cell divides twice, producing four offspring cells before the chromosomes can replicate themselves again. Thus, each offspring cell has only half the number of chromosomes contained in the original cell. For example, meiosis takes place in the cells that give rise to sperm and eggs. Later, when a sperm fertilizes an egg, the original chromosome number is restored, but half the genetic material comes from one parent and half from the other.

The evolution of sexual reproduction late in the Proterozoic

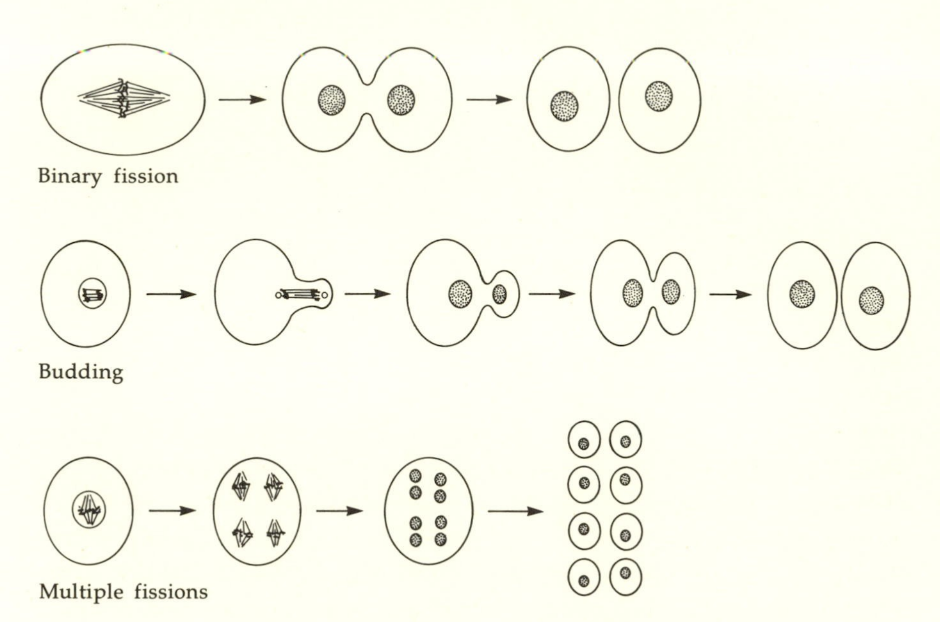

Figure 5-1. Asexual reproduction in eukaryotes. The most common mode is "binary fission," in which a cell divides into two equal parts to produce two offspring cells. Binary fission is the usual mode of prokaryotic division. Budding, which is less frequent, is common in budding bacteria and yeast. The offspring cell is smaller than the parent cell but contains the same genetic material. It gradually reaches the size of the original parent. Multiple fission is rare. The parent cell gives rise simultaneously to several or even hundreds of genetically identical smaller offspring cells.

added a major strategy to life's repertoire: it was a new means of incorporating adaptive genetic changes. Pre-existing, and therefore presumably advantageous, genes could be rapidly recombined in new individuals. Unlike the primarily asexual prokaryotes, sexual organisms need not wait for the occurrence of a favorable mutation in an organism already carrying other favorable mutations. Different individuals carrying favorable traits can bring them together quickly. With the introduction of sexual reproduction, genetic rearrangements took place in every generation. There were more gene combinations for selective factors in the environment to act on. As a result, a new sort of evolution supplemented the long period of microbial metabolic evolution.

Changes in form and size gave rise to protoctists ancestral to fungi, animals, and plants. The era of diversification of large organisms began.

Chromosomes and Mitosis

In prokaryotes, the genetic material is a circle consisting of a single long DNA molecule. The far larger eukaryotes—containing mitochondria, plastids, undulipodia, and other organelles—have more genetic material (see Table 5-1). Most eukaryotes have about a thousand times more DNA per cell than prokaryotes have (some have as much as 10,000 times more), although they have far fewer than a thousand times as many genes. Not all the reasons for the extra DNA are known. Certainly the quantity of DNA is related—although not in a simple way—to the size and capabilities of the organism carrying it.

In a multicellular eukaryote, such as an animal or plant, every cell carries the entire complement of genes of the whole organism. However, the cells in an animal or a plant evidently are not all alike. The various genes do not act unless they are signalled to do so. Switching on a cell that secretes a specialized protein, for example, or signalling a cell to cease dividing or to become a nerve cell rather than a muscle cell, is accomplished by intricate mechanisms still poorly understood. Even so, it seems clear that the development of such complex control systems for gene expression must have required increased quantities of genetic material. Extra DNA apparently is also required for the chromosomes to "dance," to pair and segregate in an orderly way in meiosis. The very fact that eukaryotes continue to replicate DNA, in great quantities, that is not needed to synthesize proteins suggests either that they have much greater need of it than prokaryotes do, or that the DNA—for selfish reasons—has managed to persist in these cells.

Chromosomes consist of long DNA molecules coiled and recoiled into compact rod-shaped bodies inside the cell nucleus (see Figure 5-2). They also contain RNA, as well as several kinds of proteins (primarily, histones). In fact, by weight, chromosomes are composed mostly of protein, not DNA. With very few

Table 5-1. Amount of DNA in Various Cells and Organelles

Organism	Length of DNA in micrometers (10^{-6} meter)
BACTERIA	
Escherichia coli (colon bacterium)	1200
Mycoplasma gallisepticum (wall-less bacterium)	300
Cyanobacteria ("blue-green algae" of various species)	500–5000
PROTOCTISTS	
Trypanosoma gambiense (sleeping-sickness parasite)	60,000
Gymnodinium (dinoflagellate)	2,300,000
Paramecium (ciliate) mitochondrion	14
Acetabularia (green algae, several species) chloroplast (more than 1,000,000 chloroplasts in each cell)	200–1000
FUNGI	
Saccharomyces cerevisiae (yeast)	6000
S. cerevisiae mitochondrion (several strains)	6–25
ANIMALS	
Rana pipiens (frog)	2,300,000
R. pipiens mitochondrion	6
Homo sapiens	1,000,000
H. sapiens mitochondrion	5
PLANTS	
Ranunculus (buttercup)	500,000–17,000,000
Pisum sativum (pea) mitochondrion	30
P. sativum chloroplast	40

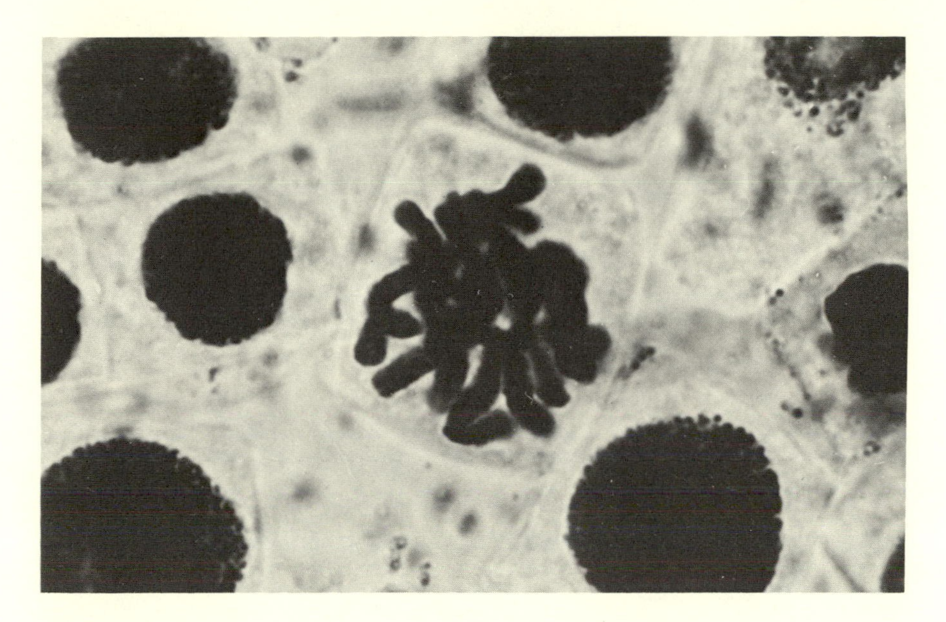

Figure 5-2. A cross section of the tip of an onion root. The cell in the center has begun to divide—the DNA and histone protein in its nucleus have just condensed into chromosomes. The large dark spheres elsewhere in the picture are the nuclei of cells that are not in mitosis—their threads of DNA and protein are diffused throughout the nucleus. The cell is about 10 microns in diameter.

exceptions, chromosomes are visible only while a cell is dividing or preparing to divide. At other times, the cell's DNA is uncoiled and dispersed throughout the nucleus, which then appears, under the light microscope, to be a relatively featureless body.

All eukaryotic cells, no matter how primitive, have at least two chromosomes. However, the number may vary enormously: there are two in each cell of the dandelionlike plant *Haplopappus ravenii*; *Aulacantha scohymantha*, a spiny and beautiful marine protist belonging to the actinopod group, has approximately 1600 chromosomes in its single cell. Human beings have 46 chromosomes (23 pairs) per cell, and the chimpanzees, probably our closest living relatives, have 44.

Mitosis is the kind of cell division by which single-celled eukaryotes reproduce without sex. It is also the means by which multicellular eukaryotes grow from a single cell (such as a fertilized egg) or regenerate damaged tissue. In mitosis, a cell divides into two offpsring cells. Before it divides, the parent cell makes a copy of its chromosomes, so each of the two offspring cells receives a set of chromosomes identical to the set in the parent cell. No matter how many chromosomes a cell has, mitosis assures the precise distribution of the chromosome sets to the two offspring cells (see Figure 5-3).

Mitosis begins when the dispersed DNA in a cell's nucleus coils up and condenses into chromosomes; the membrane enveloping the nucleus begins to disappear. The cell has already replicated its DNA, and the chromosomes can be seen to be double—each consists of two identical rods joined somewhere along their length by a small disk or cup-and-ball indentation, called a centromere or kinetochore. As time passes, the mitotic spindle appears, consisting of microtubules that radiate from opposite sides, or poles, of the cell. The kinetochore of each chromosome becomes attached to at least one of these microtubules and is maneuvered into the plane of the cell's equator, midway be-

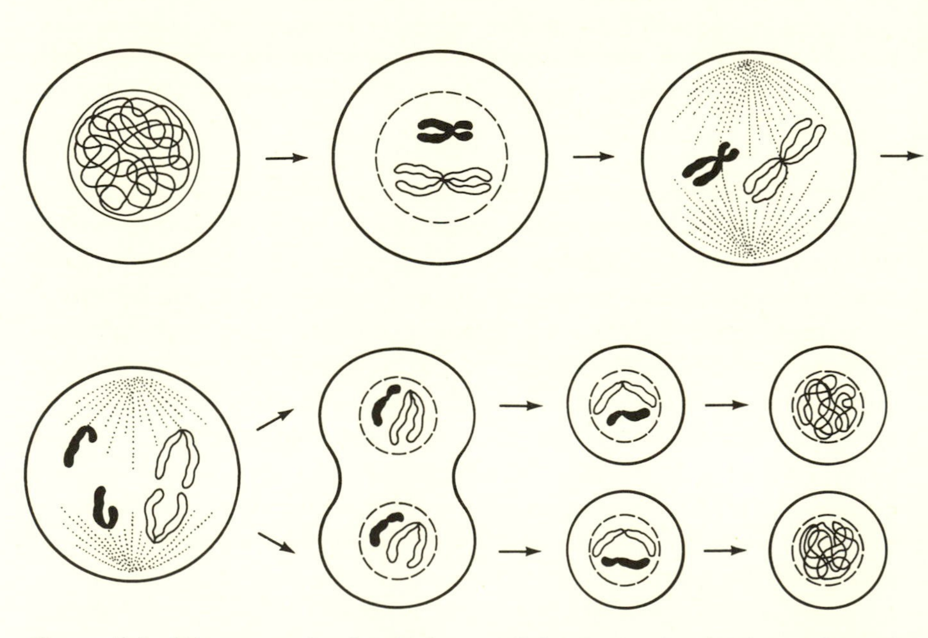

Figure 5-3. Diagram of mitosis in a cell having a haploid number of chromosomes. The cell divides into two cells genetically identical to the original.

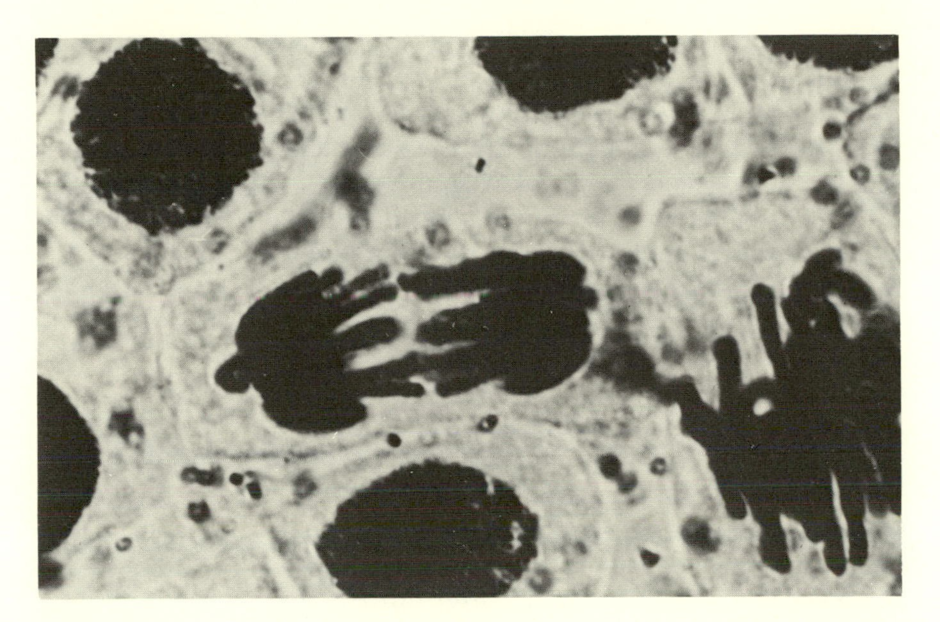

Figure 5-4. A cross section of the tip of an onion root. The center cell (diameter about 10 microns) is nearing the final stage of division—the chromosomes have divided and equal sets are moving toward opposite poles of the cell. The cell at the right is not as far along—its double chromosomes have just arranged themselves at the plane of the cell's equator.

tween the poles. The kinetochores then divide, and the two rods forming each chromosome separate and proceed toward opposite poles of the cell (see Figure 5-4). When they reach the poles, they begin to uncoil, the mitotic spindle disappears, and a new nuclear membrane forms around each of the two sets of chromosomes. Division is completed when a constriction around the equator pinches the cell into two parts (in animals) or a new stretch of cell wall forms in the plane of the equator (in plants).

The Mitotic Spindle

Once mitosis and cell division are complete, the mitotic spindle disappears. The tubulin proteins making up its microtubules dissolve in the cytoplasm, from which they reappear to form microtubules at the next mitotic cell division. The means by which these changes are accomplished are not precisely known.

It is known that dissolution and re-formation of the spindle are sensitive to changes in calcium concentration, to low temperature, to high pressure, and to certain chemicals, such as colchicine. These agents can induce the spindle's premature disappearance, at which point all chromosome movement abruptly ceases. Often, a return to normal conditions will cause the spindle to re-form, whereupon chromosome movement promptly resumes. Also, a chromosome severed from its direct connection to the spindle can never move. The region where the spindle fibers attach—the kinetochore—can move even if the rest of the chromosome is severed from it (this can be done in the laboratory by irradiating the cell with ultraviolet light).

In addition to the spindle, other structures composed of microtubules are observed during mitosis. At the poles of the spindle in most animal cells are two centrioles, small bodies that early microscopists saw only as dots. In many organisms centrioles are always at the poles in mitosis; after mitosis they become kinetosomes for undulipodia. Each centriole replicates itself during cell division, and the new centriole migrates to the opposite side of the nucleus. Thus, at the next division, there is always a centriole in place at each pole.

Although centrioles appear to play only a passive role in mitosis, they convey to the offspring cells the potential for rapid formation of undulipodia. Centrioles have an intricate structure. They are composed of nine triplets of microtubules arranged in a circle (see Figure 5-5). In almost every detail, they are just like kinetosomes—the basal bodies of undulipodia; they only lack cilia or eukaryotic flagella growing out from them (see Figure 4-14). In fact, the ciliated cells lining animal throats and nasal passages arise from dividing unciliated cells whose centrioles then turn into kinetosomes. The similarity is provocative, suggesting a common origin for all microtubule-based structures—undulipodia, the mitotic spindle, and centrioles. My hypothesis—outlined in the previous chapter—is that spirochetes, which already contained microtubules, attached to protist hosts and became functional undulipodia. In time, some of these spirochetes were drawn inside the host cell, and their components—tubulin proteins—were transformed into mitotic spindle microtubules and centrioles.

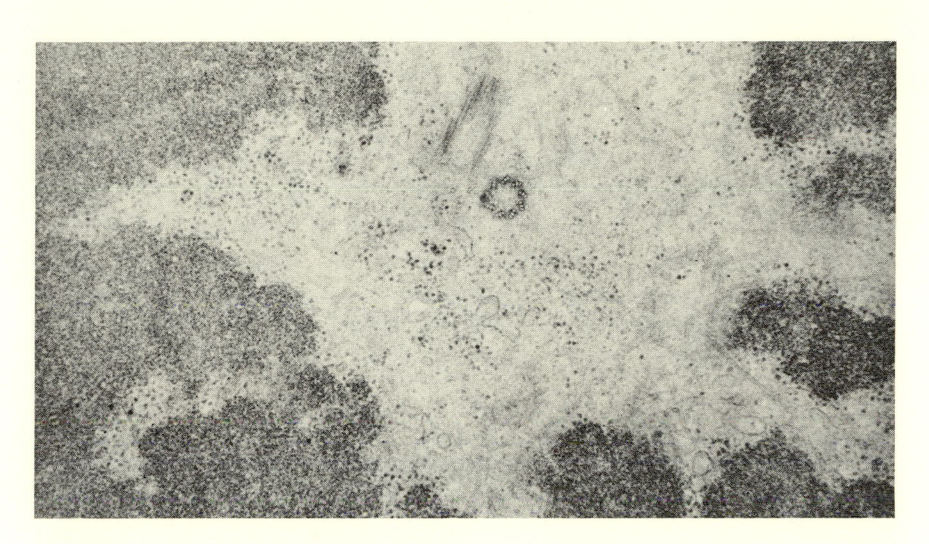

Figure 5-5. At upper center are two centrioles at right angles to each other—one in cross section (diameter about 1/4 micron) shows nine triplets of microtubules. This cell (from a Chinese hamster) has been treated with Colcemid, a chemical that dissolves the microtubules of the mitotic spindle. Thus, the centrioles are not at the poles but at the center of a circle of chromosomes, the dark forms at the edge of the picture. [Courtesy of B. R. Brinkley and E. Stubblefield, University of Texas.]

In return for mitotic reproduction, however, some protist hosts had to pay a price: they lost their motility. Once they had drawn in the microtubule system and used it for intracellular purposes, such as making mitotic structures, from an evolutionary point of view they had traveled down a blind alley. They had no further means of remaking the protruding undulipodia and thereby recovering their motility. This inability would explain why many organisms descended from these protists have well-developed mitotic systems but always lack motility by undulipodia. Such organisms include molds and mushrooms, some amoebas, and the red seaweeds.

Other eukaryotic microorganisms (for example, the amoeboflagellates) retained their motility, but only at certain stages of their life cycle, when the undulipodia reemerge and play their ancient role as oars. At other stages, these cells withdraw the undulipodia, whose microtubules are then used for mitotic processes. Several other groups of protists, such as the ancestors of some dinoflagellates and ciliates, retained motility by undulipodia along with the ability to divide. They accomplished this feat by utilizing some microtubules and related structures for the mitotic apparatus and others for the undulipodia. The ancestors of the animals, however, never achieved this duality; once an animal cell develops undulipodia, that cell loses the capacity to divide. New undulipodiated cells, such as sperm, invariably arise from cells lacking any form of undulipodia.

Where did the mitotic process come from? The regular division of chromosomes that occurs in each mitosis seems to function precisely in all animal, plant, and fungal cells. Yet there are no precedents for mitosis in prokaryotes; aside from some spirochetes, they do not even have any microtubules. One would not expect mitosis to have developed in a straight-line manner. There must have been numerous dead ends, variations, and byways.

Electron microscopic studies of single-celled eukaryotes, including amoebas, ciliates, dinoflagellates, euglenas, and green algae, have revealed some of the experiments that probably were made in the evolution of mitosis. Some asexual amoebas, for example, contain unwound DNA complexed with protein but they never form chromosomes. In some dinoflagellates, however, the chromosomal material never unravels; chromosomes remain tightly coiled even when they are not in mitosis, and typical chromosomal proteins (histones) are missing entirely. In other dinoflagellates, the coiled chromosomes are directly attached to the nuclear membrane, an arrangement reminiscent of bacterial division. In some yeasts, the spindle is so reduced in size that there is only one microtubule for each chromosome; it is not visible with a light microscope at all. In some euglenas, chromosomes seem to wander to the poles at different times; there is no orderly separation of groups of chromosomes. The refinement of mitosis may have occupied as much as a billion years of pre-Phanerozoic time. By the late Proterozoic, however, about 700 million years ago, coelenterates and various worm groups having highly developed, modern mitotic patterns are much in evidence.

Meiosis

The majority of animals, plants, and fungi have a sexual stage sometime in the life cycle. In some protists, parents of opposite sex simply fuse their entire bodies. In multicellular organisms, single specialized cells or their parts fuse, as sperm head and egg do in animals or pollen tube and embryo sac nuclei do in plants. The fusion of these reproductive cells, also called gametes, gamete nuclei, or germ cells, is known as fertilization. This results in the formation of a new cell, called a zygote, containing a set of chromosomes from each parent. Eventually, the organism that develops from the zygote produces gametes or gamete nuclei, completing the cycle.

The gametes of each eukaryotic species contain a characteristic number of chromosomes, called the haploid number. The diploid number (just twice the haploid) is the number of chromosomes in a zygote. Human zygotes, like all other human body cells, contain 46 chromosomes; the eggs and sperm, however, contain only 23 each. Evidently, some event after the formation of the zygote, but before the next round of sexual fusion, reduces the number of chromosomes in some cells back to the haploid number. That event is meiosis, a kind of cell division that leads to new cells having only half as many chromosomes as the original cell; haploid cells are formed from diploids (see Figure 5-6). Meiosis occurs in certain human cells before the development of eggs and sperm. In females, it occurs before birth, when eggs form in the ovaries of the fetus; the eggs shed by the adult ovary prior to fertilization are the products of meiotic events that occurred years before. In males, meiosis begins in the testes, at puberty, and thereafter takes place continuously to form sperm cells.

Meiotic reduction of chromosome number followed by fertilization to reestablish the chromosome number may seem an unnecessarily complicated system for producing new individuals, but it offers a significant evolutionary advantage: it is a rapid means of combining advantageous traits from different individuals. Also, genetic variation in the offspring is clearly greater when sex occurs in every generation, because no two siblings (except identical twins) receive exactly the same half of their parents' chromosome complement.

Figure 5-6. Meiosis, or reduction division, by a cell having four chromosomes. As in mitosis, each chromosome condenses already in doubled form, the two identical parts being joined at a constricted point, the kinetochore. However, there are also two similar (homologous) chromosomes of each kind—one from each parent (stage 2). The homologous chromosomes position themselves opposite each other across the cell's equatorial plane. Before cell division separates them, they can trade sections of DNA with each other (stage 4). The first cell division (stages 5 and 6) then yields two cells, each with half the number of chromosomes in the original cell. The second cell division (stages 8 and 9) is like mitosis in that it merely separates doubled chromosomes at their kinetochores. The final result is four different cells.

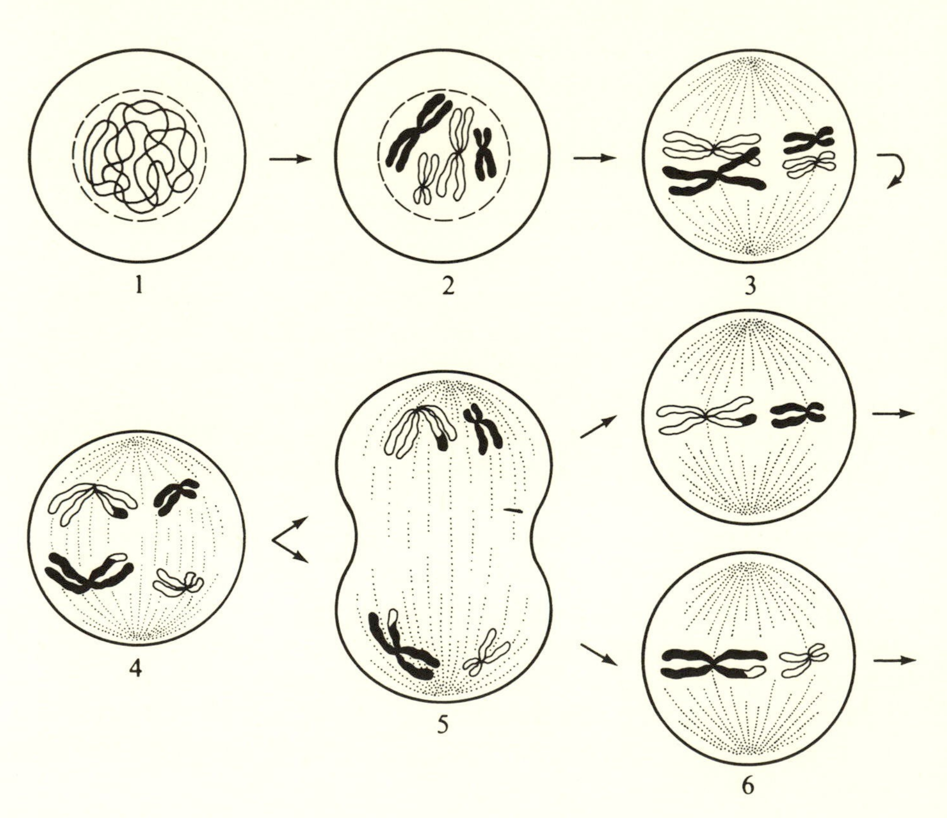

The Evolution of Meiosis

When meiosis malfunctions, chromosome imbalances can occur. Down's syndrome (mongolism) and certain intersex syndromes (genetic disorders that produce individuals intermediate in sexual constitution between male and female) result from a sperm or egg that has an extra chromosome or only an extra fragment of one.

Presumably, such genetic imbalances occurred in great number during the evolution of meiosis. Many groups of protoctists either have not evolved regular meiosis or show idiosyncratic variations that are presumably vestiges of stages in its evolution. The trypanosome mastigotes (some of which cause sleeping sickness), for example, often have intricate life cycles that involve different hosts; they undergo great changes in form when they move from one host to another, and even within a single host, yet they have no meiosis. The euglenids and certain amoebas also reproduce without meiosis. The number of chromosomes and the quantity of DNA can differ enormously between two cultures of euglenids that otherwise seem to be closely related. They can even differ between two individuals in the same species under different physiological conditions. For example, cells grown in a scarcity of phosphorus contain less

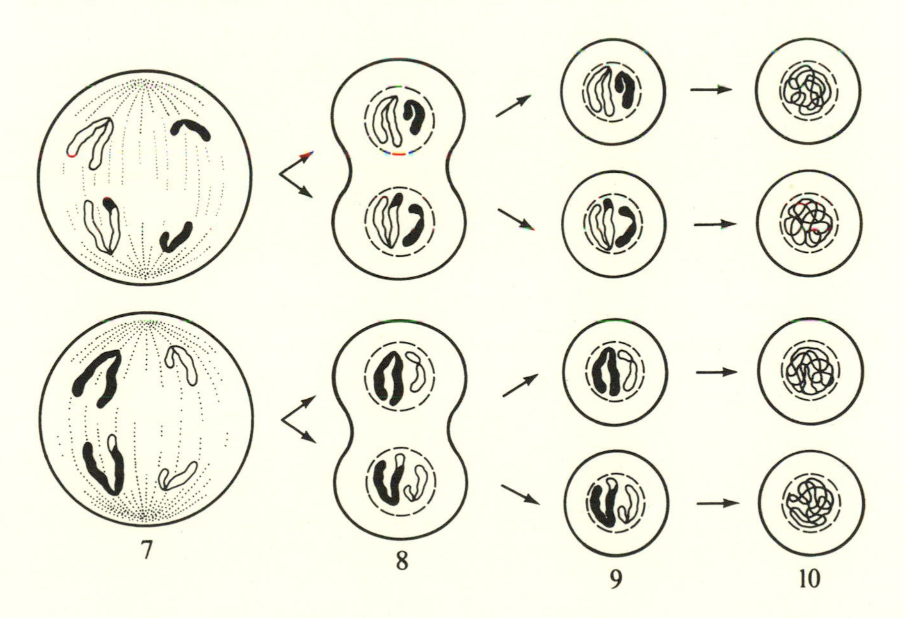

DNA. Such genetic "looseness" is not seen in more recently evolved organisms. For example, the total quality of DNA and the over-all chromosome organization in cells of closely related species of birds or mammals are highly similar.

It may have taken 500 million or a billion years for meiosis to have evolved in early eukaryotes. Or it may have taken a much shorter time. The perfected meiotic cycle may have evolved once in some long-gone single-celled ancestor of all of the animals, fungi, plants, and meiotic protoctists. I rather doubt this. More likely, selection pressures for even distribution of genetic material, fertilization, and reduction division to reestablish the single set of chromosomes were exerted on several lines of protoctists. Because meiotic divisions are just highly specialized and controlled variations on mitosis, meiosis could have arisen only in groups of organisms in which chromosomes, the microtubular spindle, and the full-fledged mitotic cycle had already evolved.

L. R. Cleveland (1898–1969), of Harvard University, observed remarkable variations on the theme of mitosis in protists found in the hindguts of wood-eating cockroaches and termites. He described a structure called the "long centriole" to which the chromosomes were attached. Cleveland explained how some of the mitotic variations could be considered analogous to stages in the evolution of meiosis. For example, a critical step in mitosis is the growth and duplication of the kinetochore, or, in the case of Cleveland's mastigotes, the "long centriole." If these structures fail to divide, the offspring chromosomes will fail to separate and do not travel to opposite sides of the cell. Instead, they all go to one pole or the other because they remain attached to the same structure. The movement of offspring chromosomes to the same pole is a malfunction in mitosis, but it is essential for the success of the meiotic process.

Failure of the kinetochore to divide occurred sporadically, as an accident, in some of Cleveland's protists. Because such accidents lead to offspring cells having too many or too few chromosomes, it was lethal to Cleveland's protists, as it probably was to many organisms in the past. However, similar meiotic irregularity, one in which the resulting offspring tolerated the still doubled chromosomes, was probably a step in the origin of meiosis. Cleveland eventually constructed a theoretical evolutionary pathway leading in a logical fashion from mitosis to meiosis and fertilization. Unfortunately, his ideas were largely ignored (and sometimes disdained) by his colleagues. Cleveland realized, although he did not clearly state it, that the evolution of meiosis from mitosis requires at least three steps: (1) the formation of a diploid cell, (2) the pairing of the homologous chromosomes of that cell, and (3) the failure of the kinetochore to divide, so that each doubled chromosome travels to one pole or the other instead of dividing into offspring chromosomes that travel to opposite poles.

Meiotic sex probably originated from cannibalism provoked by hard times. A diploid cell would be formed when one single-celled eukaryote ate a member of its own species but failed to digest it. Diploids are also formed if mitosis and nuclear division occur but cytokinesis, cytoplasmic division, fails. The resulting nuclei may fuse with each other. Cleveland observed such irregularities and I have observed cannibalistic fusions of protists (stentors, for example) in busy pond water communities. The pairing of chromosomes and the failure of kinetochores to divide probably happened casually at first, but in time were selected for. Although the new diploid cells, being larger, may have been better suited to adverse conditions, the return of former conditions would have restored the advantage to the haploids. Thus, the selection pressure for the pairing of homologous diploid chromosomes, their failure to divide, and their equal distribution to haploid offspring cells must have been relentless: after unequal distribution, at least one offspring cell would have been crippled by the absence of the genes of one or more chromosomes.

The variety of sexual systems found in living protists, such as ciliates, water molds, slime molds, and green algae, suggests that meiosis evolved in separate groups, but groups that had many fundamental features in common. *Chlamydomonas*, a useful organism for genetic studies, is a single-celled green alga that lives in fresh water and damp soil (see Figure 5-7). These undulipodiated algae swim around and function perfectly as haploid individuals, having only one set of chromosomes. They regularly reproduce asexually—a single *Chlamydomonas* can divide by mitosis into two offspring identical to the parent and to each

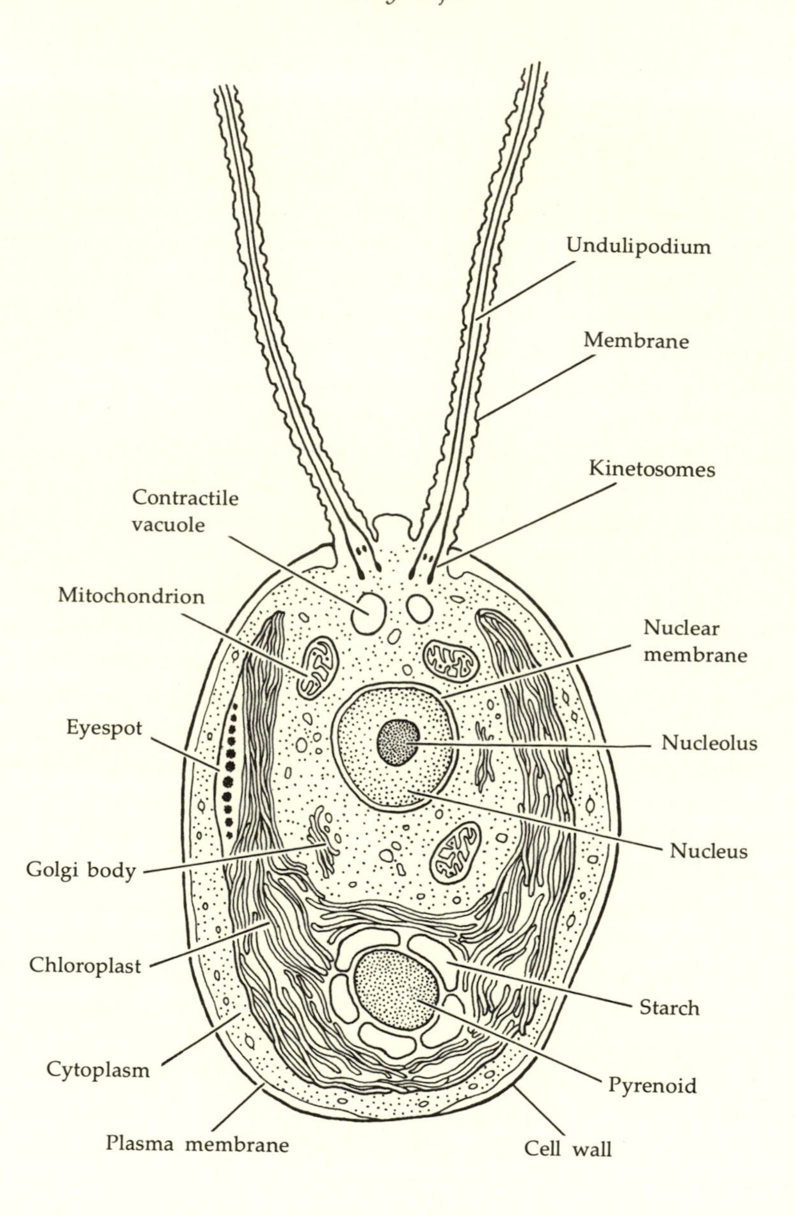

Figure 5-7. Chlamydomonas, a single-celled alga, has one large cup-shaped chloroplast. As in many other algae, the chloroplast contains a pyrenoid, a spherical body surrounded by starch granules. The pyrenoid is thought to contain enzymes that catalyze the conversion of sugar to starch, a storable form of photosynthetic products. [Drawing by Laszlo Meszoly.]

other. However, they occasionally reproduce sexually. For reasons not yet understood, nitrogen starvation induces sexual fusion in cultures of these microorganisms. If nitrogen-starved plus and minus cells (members of the same strain but different mating types) are mixed together in a tube or on a glass plate or Petri dish, they come together within minutes and fuse.

Their undulipodia, which contact first, seem to carry sex-distinguishing factors, probably proteins; if their undulipodia are removed, two cells will not enter the sexual process. (But—pathetically, perhaps—undulipodia alone, sheared from cells of opposite mating type, do recognize each other and stick together.) Because the separate algae had one set of chromosomes apiece, the zygote resulting from their fusion has two—it is a diploid cell. The zygote also has double the normal number of organelles: two chloroplasts instead of one, twice as many mitochondria as before, four undulipodia, and so on.

Immediately after fusion, the zygote enters a resting stage, the zygospore. Whatever happens next is hidden from view for about four days. The zygospore turns black on the outside, undulipodia are withdrawn, and there is no cell movement during this time.

Inside the zygospore, the diploid cell undergoes meiosis to restore the normal state—a single set of chromosomes in each of four cells. How the mitochondria and other organelles and cell materials reestablish their proper numbers and quantities is still not clear. The zygospore finally opens and four little cells swim out. Each looks like the original parents on the outside and has only one set of chromosomes, but they differ considerably in their complements of genes. Each new haploid cell has a single set of chromosomes, but not all of them are from the same parent. And in the zygote, when all the chromosomes are together, the similar chromosomes from different parents can exchange genes with each other. Thus, each chromosome in the offspring can contain parts from different parents. The four offspring are more like fraternal siblings than identical quadruplets.

In *Chlamydomonas*, favorable genetic traits—once in separate parents—may be brought together in new individuals. For example, one recombinant offspring may be better able to tolerate the scarcity of nitrogen, which induced the pairing in the first place. Other protists illustrate different variations that must have appeared during the evolution of meiosis. Once the meiotic reduction–fertilization cycle became stabilized, it was obviously of great selective advantage. All the large, multicellular eukaryotes of today have meiotic ancestors.

If vagaries of environment did not exist, and if weather, climate, and geological phenomena were predictable and regular, sex might never have evolved. In fact, certain highly successful organisms, such as sponges and a huge group of fungi (the fungi imperfecti), living in stable environments, have reverted to asexuality and produce offspring like their parent. Over most of the world, however, change is the rule and species with the capacity for sexual production of offspring are more plastic, more versatile than those that receive all their genes from a single parent. Good adaptive traits, highly successful genes selected for in populations, can easily be brought together in single individuals by sexual means. Such highly adapted individuals now carrying advantageous genes from both parents may interbreed and found new populations. With time, such new populations may diverge significantly from ancestral populations; new flexible, well-adapted species arise and replace the old ones.

By the time of the late Proterozoic, some 700 million years ago, the major genetic mechanisms for insuring variation through sexual reproduction had evolved. Meiosis and mitosis were well established in the early animals—worms, coelenterates, and arthropods—that flourished before the Cambrian. The major metabolic pathways—fermentation, photosynthesis, and respiration—had long been developed. Atmospheric gas regulation continued, still accomplished largely by interacting communities of the metabolically diverse prokaryotes. Probably the major effect eukaryotes have had on the planet is on the rates of elemental cycling, rather than on which elements are being cycled. The evolution of large organisms increased the size of the Earth's biomass and also increased the rate at which biologically essential elements were cycled. Metabolically, however, the routine was as before: oxygen released by photosynthesizers was taken in and respired in the oxidation of foodstuffs by heterotrophs.

In the organization of the cell, all the major evolutionary innovations had occurred before the beginning of the Cambrian. The stage was set for the appearance of large organisms: new invertebrates, algae, and fungi; fish, dinosaurs, birds, and mammals; cone-bearing and seed-bearing plants; large forests and grasslands. Organisms developed the evolutionary adaptations that enabled them to leave their aquatic environment and adapt to life on land. None of these achievements would have been possible without three billion years of preparation.

Suggested Reading

Cleveland, L. R. "The origin and evolution of meiosis." *Science* 105:287–288; 1947.

Grell, K. G. *Protozoology.* Berlin and Heidelberg: Springer Verlag, 1973.

Strickberger, M. W. *Genetics,* 3rd ed. New York: Macmillan, 1981.

Wolfe, S. L. *Biology of the Cell,* 2nd ed. Belmont, Calif.: Wadsworth, 1981.

CHAPTER 6

The Modern Era

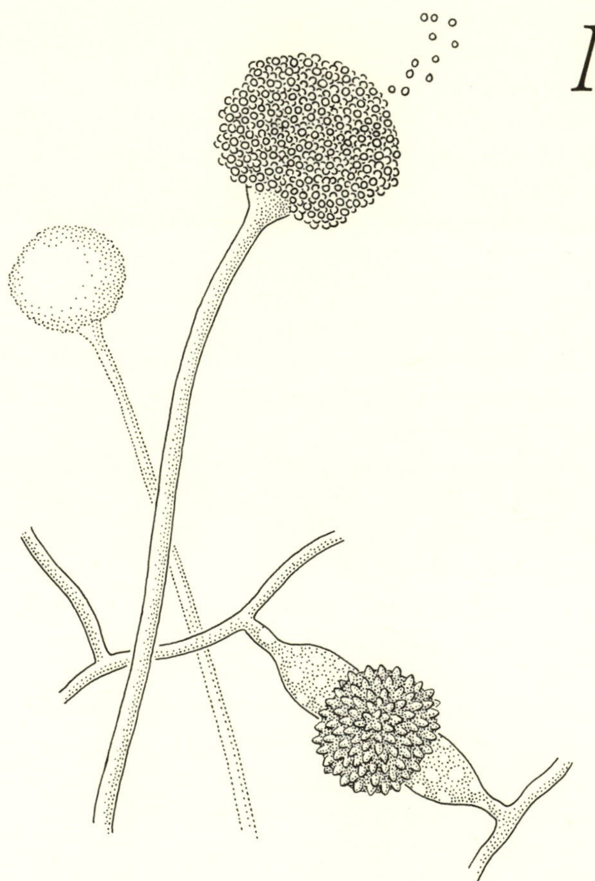

THE CONSPICUOUS organisms in our environment are made up of many cells in intricate arrangement—they are multicellular. Although there are multicellular differentiated organisms in all five kingdoms (remember Figure 1-2), the level of complexity in form is far greater in animals and plants. In bacteria, protoctists, and fungi, multicellularity is simple. Simple thready substructure is typical of fungi. Molds and mushrooms are composed of mats of similar threads in which the fungal cells are nearly identical to each other. Animals and plants, however, have many different kinds and sizes of cells. In animals, these differentiated cells make up tissues such as muscle or bone. Tissues are organized into organs—gills, blood vessels, intestines, bones, kidneys, and heart. Organs that function together, such as the skeleton or circulatory system, make up organ systems. Plants, too, have tissues organized into organs and organ systems, such as the phloem and xylem, the vascular system of flowering plants.

What innovations permitted the origin of large, differentiated organisms? How did elaborate communities of invertebrate animals, which eventually gave rise to coral reefs, emerge along the world's tropical shores? How did forests with their myriad inhabitants arise? How did the skies become populated by insects, birds, and bats? What prepared the microbial world for emergence of the modern era, the spreading of large forms of life from the ocean abyss to the mountain tops, from regions of equatorial heat to arctic freeze?

Certainly, the evolution of greater complexity in organisms required larger amounts of genetic material to act as the blueprints for increased capability. The innovation of a mechanism (chromosomes) for efficient packaging of the hereditary blueprints as well as an efficient means (mitosis) of distributing replicas to daughter cells permitted an evolutionary trend toward larger and more specialized organisms. Likewise, the evolution of sexual reproduction, with the rapid genetic rearrangements such partnerships permitted, hastened the emergence of complex forms and played a role in the diversification and spread of the new kinds of life.

Clones, Colonies, and Differentiation

There is a practical limit to the size of a single cell—and hence, the size of a single-celled organism. Cells need to interact with their environment to take in materials and to expel wastes. Many reactions vital to a cell take place at its outer membrane. Small cells are usually more efficient at this than large ones, and a unicellular organism faces serious physical constraints. Although the volume of a cell increases as the cube of its diameter, the surface area increases only as the square of its diameter. Thus, a ten-fold increase in cell diameter gives a cell having a hundred times the surface area and a thousand times the volume of the original. As a cell enlarges, the proportion of cell in contact with its outer membrane and the outside environment declines.

Some organisms have compensated for the size and efficiency problem by modifying their shape. For example, multiple infoldings of the membrane increase the proportion of surface area to volume. However, modifying the shape of a cell has its limitations. To solve this problem, vast numbers of organisms evolved multicellularity.

Single-celled organisms all divide to form copies of themselves, called clones. Sometimes the members of a group of clones fail to disperse. This failure can be put to advantage. Cells at the exterior of the clone group have more exposed surface so their environment is different from the environment of the cells in the interior. There is a tendency for the clones to differentiate into a colony, a group of organisms all deriving from the same parent but differing one from another. Natural selection acts to preserve the members of the clone as a functioning unit. The conversion of clones to colonies and then to differentiated multicellular individuals occurred in hundreds of ancestral lineages—all the various protoctist and fungal forms and certainly in those that were ancestors to animals and plants. In all these lineages, the individual cell remains small (and efficient), but the organism is composed of numerous cells. Furthermore, just as organelles within a cell assume particular functions, groups of cells in a multicellular organism differentiate during development to take on specialized functions for the organism. Thus, multicellularity made for efficiency and the potential for spectacular diversification of body form.

When did the ancestors of animals and plants first make the evolutionary leap from unicellular existence to multicellularity? Unfortunately, there is no clear fossil record for periods earlier than the Cambrian (about half a billion years ago). Therefore, the story of the origin of animal and plant multicellularity must draw heavily on the comparative study of existing organisms.

The volvocines, a group of colonial algae, illustrate a possible route from single-celled organisms through clusters of independent cells to organisms made of interdependent cells (see Figure 6-1). The individual cells of the volvocines closely resemble *Chlamydomonas* cells—each has two undulipodia, an eyespot, and a chloroplast (see Figure 5-7). *Gonium*, the simplest of the colonial volvocines, is a tiny disk of 4, 8, 16, or 32 such cells (the number depends on the species) held together by gelatinous substances. All the undulipodia protrude from the same side of the disk, and their beating pulls it through the water. Each cell of a *Gonium* colony is able to divide to produce a whole new colony. *Pandorina* consists of 16 or 32 cells (depending on the species) packed together in an egg-shaped ball. The eyespots of the cells at one end of the colony are larger than those of the other cells, and this defines a forward direction for the colony as it spins through the water like a football, propelled by the beating undulipodia. Each cell of a *Pandorina* colony, too, can produce a new colony.

The most complex of the volvocines is *Volvox* itself, a great hollow sphere made of a single layer of from 500 to 600,000 cells (depending on the species). Each sphere has forward and rear poles, and it rotates on its axis as it moves forward. Only a few of

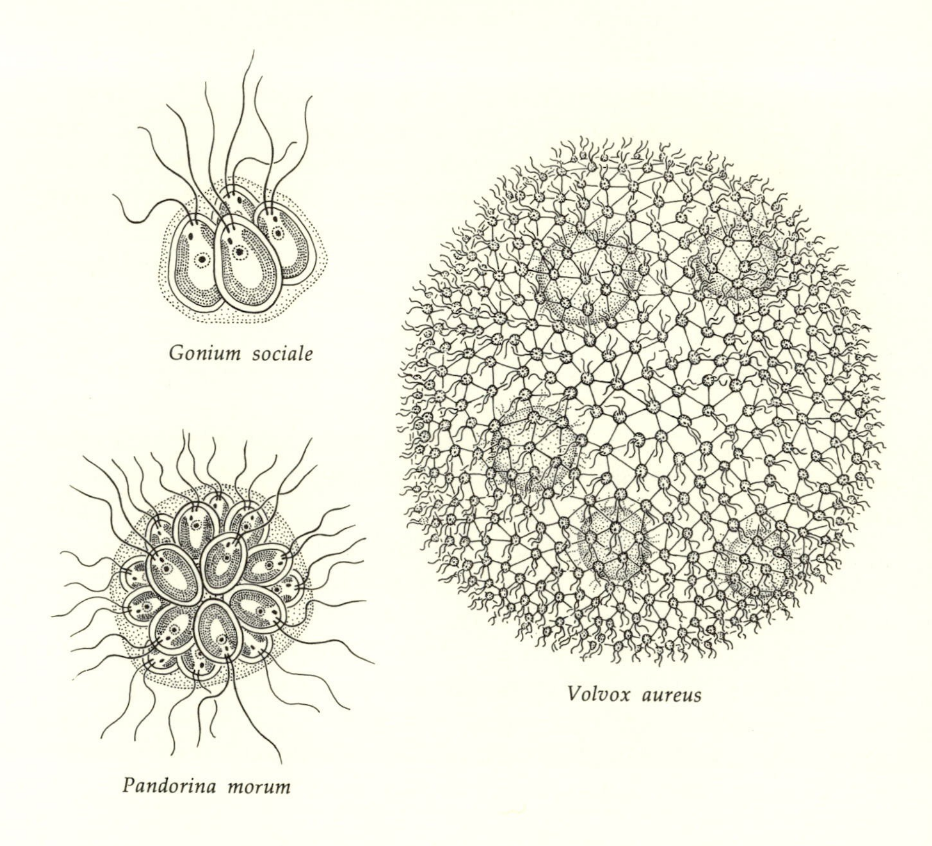

Figure 6-1. Three volvocines. [Drawings by Laszlo Meszoly.]

the cells can divide to produce offspring colonies, and these cells at first are typically near the rear pole of the sphere. Later, each divides repeatedly to produce offspring spheres floating in the hollow interior of the parent colony; eventually, these release an enzyme that dissolves the gelatin holding the parent sphere together—they hatch. *Volvox* can also reproduce sexually. In some species, each sphere can produce both eggs and sperm; in others, a given sphere may produce either eggs or sperm, but not both; in still another species, each sphere has a genetically determined sex. *Volvox* has crossed the threshold between coloniality and true multicellularity. Most of its cells cannot live independently. In the most differentiated species, reproductive cells have lost their undulipodia and are immotile. They have assumed the specialized function of reproduction while other cells carry out the function of locomotion.

Another clue to the evolution of multicellularity may be seen in the early stages of development of modern animals. Virtually all living metazoans (animals that form from embryos and are composed of tissues and organs) share certain characteristic features of early development: the fertilized egg (zygote) divides several times and forms a hollow ball of cells, the blastula. *Trichoplax adhaerens* is a tiny amoebalike marine animal that resembles a flattened blastula (see Figure 6-2). Although clearly multicellular, it lacks tissue development and shows little specialization of function. Because blastula formation is so characteristic of metazoans, *Trichoplax* may be similar to the early, primitive animals of the Proterozoic.

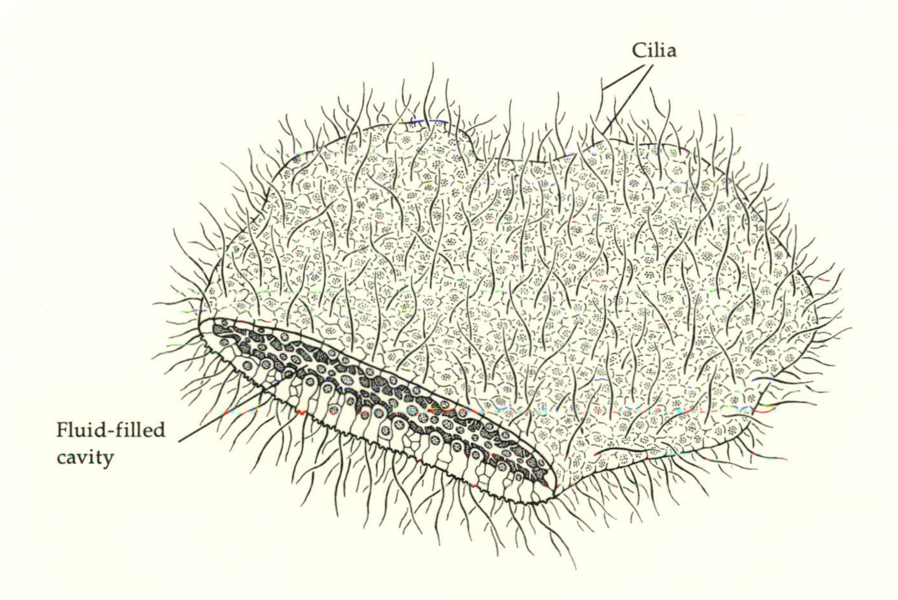

Figure 6-2. Trichoplax adhaerens was first discovered crawling on the glass of a marine aquarium. An adult *Trichoplax* is about half a millimeter wide. Its shape is as simple an animal as can be imagined—a flattened ball of ciliated cells. Although the top differs from the bottom in having fewer cilia, *Trichoplax* has no symmetry, no front or back, right or left—it can crawl in any direction. [Drawing by Laszlo Meszoly.]

The Cells of Multicellular Organisms

Whatever the pathways to multicellularity, its chief evolutionary advantage lies in the capacity for differentiated function. The sheer number of cells permits individual and group specializations not possible to single cells or to small colonies of them.

Like meiosis, multicellularity evolved many times in different groups of organisms. The first step in many cases, as for instance in fungi and red seaweeds, was the failure of the offspring of single-celled parents to separate after division. Thus, long strands or filaments formed. As in the weaving of threads to make cloth, such filaments became matted and intertwined, forming flat leaflike structures, or else they built up three-dimensional cell colonies. The subsequent specialization of groups of cells created a need for different kinds of intercellular connections. In animals, such connections are extremely elaborate, far more complex than the connections in plants and fungi. In some tissues, such as connective tissue, adjacent cells are separated from each other but are held together by a coating or matrix secreted by the cells. In many tissues, however, the cells must communicate with each other, and they do so through openings of various sizes in the membranes of adjacent cells. Such connections allow multicellular organisms to coordinate the activities of their cells in embryonic development, differentiation, growth, and metabolism. Small openings allow ions and small organic molecules to pass but block the passage of large molecules. In some plant tissues, adjacent cells are connected by cytoplasmic channels that allow large molecules and even some organelles to flow between the cells.

Multicellular organisms also required new structural materials for support and protection. Structural innovation was particularly critical when life stepped out from the oceans and established a terrestrial foothold. Both plants and animals needed physical support to compensate for the loss of buoyancy and protection from water loss by evaporation. The success of this transition and the rapid proliferation of diverse life forms was tied to the ability of cells to secrete nonliving, extracellular material for protection and support.

Nearly all cells have some kind of exterior coating outside the cell membrane. In animal cells, the coating appears to be a protein–polysaccharide complex that helps to cement adjacent cells together. Many algae and all plant cells have rigid cell walls made of cellulose. Bacteria and fungi also have rigid cell walls constructed from a complex substance called peptidoglycan, which is both protein- and carbohydrate-like. The bone of vertebrates is a matrix of the protein collagen and mineral deposits, including calcium and magnesium phosphate and carbonate. Invertebrates evolved a tough exoskeleton made up of a rigid polymer called chitin. Most terrestrial plants are covered by a thin layer of a waxy substance (cutin) that reduces water loss. Plants also produce lignin, the woody substance that (with cellulose) provides support.

The seeds of this strategy—the use of nonliving extracellular material for support and protection—had been sown in the Proterozoic. For example, in order to deposit a calcium carbonate shell, a cell must regulate the concentration of calcium in its interior. Calcium regulation probably evolved in early protists because low intracellular calcium concentration is required for the formation of the microtubules of the mitotic spindle. The concentration of calcium in seawater is some 10,000 times what it is in a cell. Thus, if seawater flows into the cell, it can swamp several processes, such as the formation of microtubules. Proteins such as that shown in Figure 6-3 have evolved to grasp the calcium and prevent it from circulating freely in the cell. In a much lower concentration of calcium, as in fresh water, the protein changes form and releases calcium into the environment. Such calcium-binding proteins were needed during the evolution of mitosis and meiosis, and later for the formation of muscles and nerves. Cells that had evolved calcium-sequestering abilities later probably evolved surface calcium-carbonate-binding proteins as dumping places for the excess intracellular calcium. Together, these developments led to the evolution of shelly hardparts. Furthermore, Proterozoic radiolarians and silicoflagellates (single-celled organisms) evolved the ability to deposit silica (silicon dioxide) in the form of shells and spines. The intracellular control of silica precipitation is not well understood, and silica-regulating proteins have not been found.

The appearance of metazoans with skeletons of calcium carbonate or calcium phosphate marks the start of modern times, the Phanerozoic Aeon. Reconstruction of evolutionary events be-

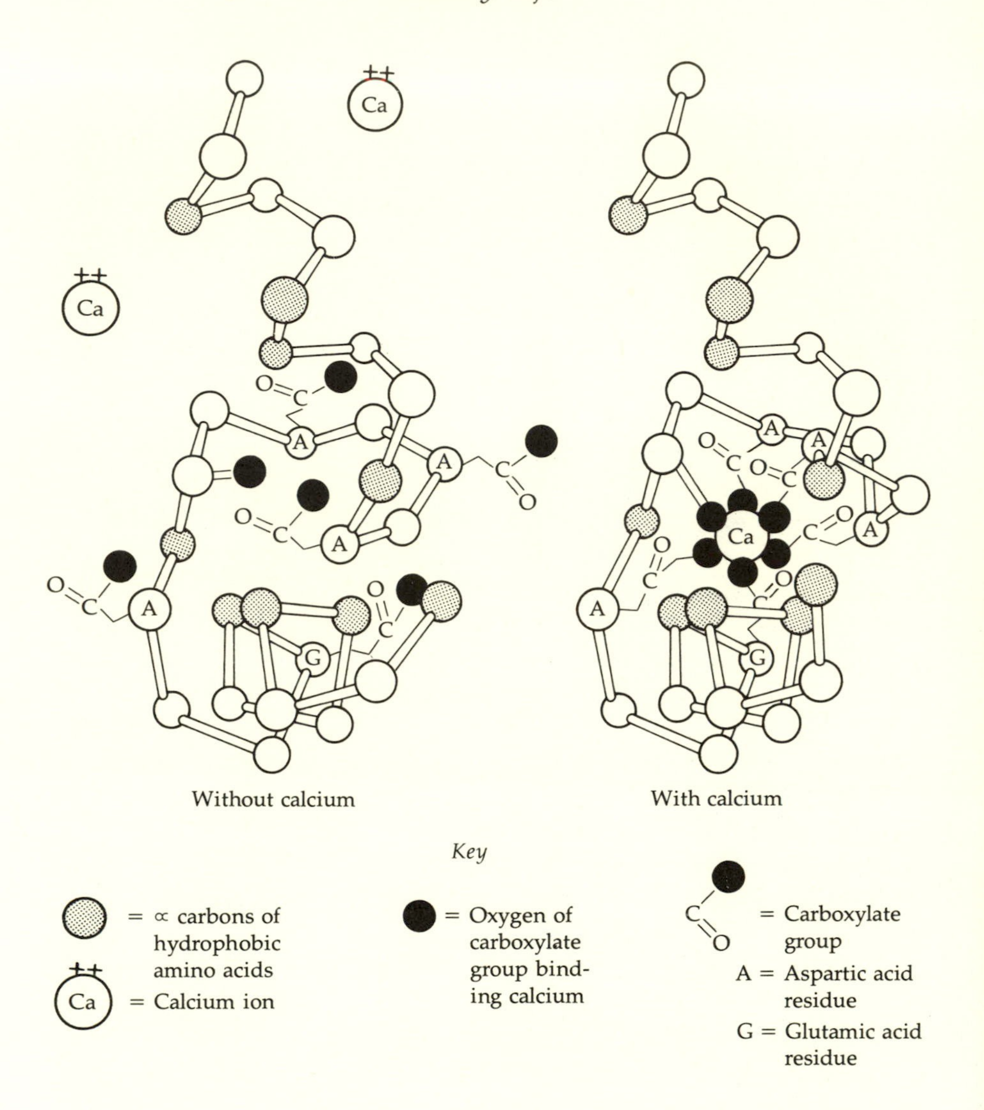

Figure 6-3. A calcium-modulating protein. A portion of the EF hand segment of the α helix is shown. On the right the calcium ion is bound by six different amino acids in the sequence. The loop of amino acid residues binds the calcium ion, stabilizing the protein. The calcium is released on the left, where the mutually repulsive carboxylic acid groups of the loop swing away from each other. This sort of calcium-modulating activity has been found in muscle, blood, and other proteins that had never been thought to be related to each other. [Courtesy of R. H. Kretsinger, University of Virginia.]

comes a simpler and more confident endeavor as fossils become more abundant and as the microbial fossil record is supplemented by one containing abundant remains of visible organisms. The major invasion of land by large multicelled eukaryotes did not take place until the late Paleozoic Era, 350 million years ago. Several now-extinct plant groups, the earliest vascular plants and their associated root fungi, came out on land first and eventually evolved into forms that could cope with total withdrawal from the water. They were followed by ferns, lycopods (similar to modern club mosses), horsetails, rushes, and gymnosperms (cone-bearing plants). The ready supply of land plants that could be eaten was a preadaptation to the proliferation of insects, amphibians, and early reptiles that took place during the Carboniferous Period.

Adaptations of many kinds enable organisms to live virtually everywhere on the Earth's surface. It is ironic that biologists first were able to confirm the ubiquity of life on Earth during preflight testing of instruments designed to search for life on Mars. In 1976, while most of the world waited to learn whether life existed on Mars, a few biologists were busy studying the results of the Viking lander's dry runs on Earth. Samples had been taken from the virtually waterless dry valleys of Antarctica, from barren lava fields on Surtsey, the new island off Iceland, and from the surface of Arctic glacial ice. Everywhere, life was found. Of course, life is most abundant and diversified in wet, rich, tropical places. As the depths of the ocean, the summits of high mountains, and the poles of the Earth are approached, biological diversity and the number of individuals tend to decrease, but never to zero.

The more that is learned about the Earth, the clearer it is that our planet's surface has been highly altered by the origin, evolution, and growth of life on it. As life expands, it alters the composition, temperature, and chemical nature of the atmosphere and the composition, texture, and diversity of the Earth's surface. The surface environment and the organisms on it have been evolving together for billions of years. My narrative has traced the evolution of cells that became structurally and functionally more intricate and that gave rise to many groups of larger and more elaborate organisms, but it would be a misreading of the evolutionary record to think of these events as a kind of upward progression.

Some authors have claimed that evolution has been "progressive," leading to "higher" and therefore better life forms. One must realize that, even three billion years ago, neatly functioning atmospheric cycles were modulated by organisms. Some two billion years ago, cyanobacteria made drastic changes in the atmosphere. It is doubtful that any organisms since then have had such a profound effect on the planet. If the vast stretch of pre-Phanerozoic time once seemed uneventful, it was because we lacked the tools to examine it. We now realize that it was the age of prokaryotic microbes. Without their achievements—their adaptations to extreme environments, their exchanges with the atmosphere, and their production of oxygen—the spectacular spread of eukaryotes would never have been possible. Without the prokaryotes' continuing activities, neither we nor the animals and plants on which we directly depend would continue to exist.

We consider naive the early Darwinian view of "nature red in tooth and claw." Now we see ourselves as products of cellular cooperation—of cells built up from other cells. Partnerships between cells once foreign and even enemies to each other are at the very roots of our being. They are the basis of the continually outward expansion of life on Earth.

Suggested Reading

Ayala, F. J., and J. W. Valentine. *Evolving: The Theory and Processes of Organic Evolution.* Menlo Park, Calif.: Benjamin/Cummings, 1979.

Hanson, E. D. *The Origin and Early Evolution of Animals.* Middletown, Conn.: Wesleyan Univ. Press, 1977.

Raup, D. M., and S. M. Stanley. *Principles of Paleontology*, 2nd Ed. San Francisco: W. H. Freeman and Co., 1978.

Glossary

actin One of the two major proteins of muscle, in which it makes up the thin filaments; also found in filaments that particpate in the motility of protists

actinopod Pseudopod containing filaments and microtubules; also, a protist that has actinopods

aerobic Requiring gaseous oxygen

anaerobic Requiring the absence of gaseous oxygen

animals Member of the kingdom Animalia, heterotrophic eukaryotes that form from hollow-ball (blastula) embryos that develop from fertilized eggs

antibiotic Substance produced by organisms, typically by bacteria, that injures other organisms or prevents them from growing

asexual Of development or reproduction in which the offspring has a single parent

ATP Adenosine triphosphate; molecule that is the primary energy carrier for cell metabolism and motility

autotroph An organism that grows and synthesizes organic compounds from inorganic compounds by using energy from sunlight or from oxidation of inorganic compounds

basal apparatus, basal body Kinetosome; also, thin cylindrical plate found at the base of bacterial flagellum

blastula Animal embryo after cleavage and before gastrulation; usually a hollow sphere, the walls of which are composed of a single layer of cells

budding Asexual reproduction by outgrowth of a bud from a parent cell or body

carotenoid Member of a group of red, orange, and yellow hydrocarbon pigments found in plastids

cell wall Structure external to the plasma membrane produced by cells; generally rigid and composed primarily of cellulose and lignin in plants, chitin in fungi, and peptidoglycans (networks of amino acids and sugar molecules) in bacteria; it is absent or of various composition in protoctists

cellulose Polysaccharide made of glucose units; chief constituent of the cell wall in all plants and green algae

centriole Small barrel-shaped organelle seen at each pole of the spindle formed during cell division; almost identical in cross section to a kinetosome

centromere *See* kinetochore

chlorophyll Green pigment responsible for absorption of visible light in photosynthetic organisms; there are several kinds, but all are organic ring compounds containing magnesium atoms

chloroplast Green plastid, a membrane-bounded photosynthetic organelle containing chlorophylls *a* and *b*

chromatid One of the duplicates of each chromosome that appear, still attached by a kinetochore, as a cell prepares to divide; chromatids separate in mitosis and move toward opposite poles of the cell

chromatin Material of which chromosomes are composed; made of nucleic acids and protein

chromosome Intranuclear organelle made of chromatin; visible during cell division; a cell's chromosomes contain most of its genetic material

chrysoplast Yellow plastid; the membrane-bounded photosynthetic organelle of chrysophytes

cilium Short undulipodium; an intracellular but protruding organelle of motility

colony Cells or organisms living together in permanent but loose association

conjugation The transmission of genetic material from a donor to a recipient cell; the fusion of nonundulipodiated gametes or gamete nuclei

cyanobacterium Synonym for blue-green alga, cyanophyte, or blue-green bacterium; member of a large group (some 10,000 species) of oxygen-producing photosynthetic prokaryotes, all obligate photoautotrophes and all containing chlorophyll *a* and other pigments called phycobilins

cytochrome Small protein that contains iron heme and acts as electron carrier in respiration and photosynthesis

cytokinesis Division of cell cytoplasm; distinguished from karyokinesis, division of a cell nucleus

cytoplasm In a cell, the fluid, ribosome-filled portion exterior to the nucleus or nucleoid

diploid Of cells in which the nucleus contains two sets of chromosomes

DNA Deoxyribonucleic acid; a long molecule composed of nucleotides in a linear order that constitutes the genetic information of cells; capable of replicating itself and of causing the synthesis of RNA

embryo Early developmental stage of a multicellular organism; produced from a zygote, or fertilized egg

endoplasmic reticulum Extensive system of membranes inside eukaryotic cells; called rough if coated with ribosomes, called smooth if not

eukaryote A cell having a membrane-bounded nucleus, organelles such as mitochondria and plastids, and several chromosomes in which the DNA is coated with histone proteins

fermentation Anaerobic respiration; the degradation of organic compounds in the absence of oxygen, yielding energy and organic end products; organic compounds are the terminal electron acceptors in all fermentations

fertilization Fusion of two haploid cells, gametes, or gamete nuclei to form a diploid zygote

flagellum (1) Long thin solid extracellular organelle of bacterial motility; composed of a protein, flagellin (2) Undulipodium; a long, fine, intrinsically motile intracellular structure used for locomotion or feeding; covered by plasma membrane and underlain by regular array of nine doublet microtubules and two central microtubules composed of tubulin, dynein, and other proteins, not flagellin. The term undulipodium refers to both the flagella and the cilia of eukaryotes; flagella are longer than cilia but have the same internal structure

fruiting body Structures that contain or bear cysts, spores, or other generative structures

fungus Member of the kingdom Fungi, heterotrophic sexual eukaryotes that reproduce from spores; they include molds, mushrooms, and lichens

gamete Mature haploid reproductive cell whose nucleus fuses with that of another gamete of an opposite sex to form a zygote

glycolysis Metabolic pathway in which glucose is broken down into organic acids and CO_2, releasing energy

Golgi body A layered, cuplike organelle composed of modified endoplasmic reticulum; plays a part in the production and storage of metabolic products

haploid Of cells in which the nucleus contains only one set of chromosomes

hemoglobin Iron-containing protein used for the transport or storage of oxygen; found in the blood of animals, the root nodules of some legumes that have symbiotic bacteria, and in some strains of *Paramecium* and *Tetrahymena* (protists)

heterotroph Organism that obtains carbon and energy from organic compounds ultimately produced by autotrophs; they include saprobes, parasites, and carnivores

histones Class of positively-charged chromosomal proteins that bind to DNA; they are rich in lysine and arginine

homologous Of structures or behaviors that have evolved from a common ancestor, even if the structures or behaviors have diverged in form and function

host Organism that provides nutrition or lodging for symbionts or parasites

hypermastigote Motile heterotroph with up to many thousands of undulipodia

kinetochore A proteinaceous structure at a constricted region of a chromosome; holds sister chromatids together and is the site of attachment of the microtubules forming the spindle fibers during cell division; also called centromere

kinetosome Organelle at the base of all undulipodia and responsible for their formation; like a centriole, its cross section shows a characteristic circle of nine triplets of microtubles

lipid One of a class of organic compounds, soluble in organic and not aqueous solvents; they include fats, waxes, steroids, phospholipids, carotenoids, and xanthophylls

mastigote Eukaryotic microorganism motile by undulipodia (a eukaryotic "flagellate") or a prokaryote motile by flagella

meiosis One or two nuclear divisions in which the number of chromosomes is reduced by half

mesosome Membranous structure associated with DNA segregation in dividing bacteria

messenger RNA (mRNA) RNA produced from DNA-directed polymerization in pieces just large enough to carry the information for the synthesis of one or several proteins

metabolism Sum of enzyme-catalyzed chemical reaction sequences occurring in cells and organisms

metamorphosis An abrupt transition from the immature to an intermediate or adult form (for example, tadpole to frog)

microbe Microscopic organism

microbial mat Matlike community of microorganisms; living precursor of a stromatolite

microtubule Slender, hollow proteinaceous intracellular structure; most are 24 nm in diameter; found in mitotic spindles, undulipodia, nerve cell processes (axons and dendrites), and other intracellular structures; often, their formation can be inhibited by colchicine

mitochondrion Organelle in which the chemical energy in reduced organic compounds (food molecules) is transferred to ATP molecules by oxygen-requiring respiration

mitosis Nuclear division in which attached chromatids (pairs of duplicate chromosomes) move to the equatorial plane of the nucleus, separate at their kinetochores, and form two separate, identical groups; subsequent division of the cell will thus produce two identical offspring cells

mitotic spindle Microtubular structure formed during mitosis and responsible for the poleward movement of the chromosomes

moneran Member of the kingdom Monera, autotrophic and heterotrophic prokaryotes that include cyanobacteria and all other bacteria

myosin One of the two major proteins of muscle, in which it makes up the thick filaments; also found in filaments that participate in the motility of protists

nitrogen fixation Incorporation of atmospheric nitrogen into organic nitrogen compounds; requires nitrogenase

nitrogenase Enzyme complex containing iron and molybdenum; converts atmospheric nitrogen to organic nitrogen

nucleoid The DNA-containing structure of prokaryotic cells; not bounded by membrane

nucleolus Structure in the cell nucleus; contains DNA, RNA, protein, precursors of ribosomes

nucleotide Single unit of nucleic acid; composed of an organic nitrogenous base, deoxyribose or ribose sugar, and phosphate

nucleus Large membrane-bounded organelle that contains most of a cell's genetic information in the form of DNA

obligate anaerobe Organism that can survive and grow only in the absence of gaseous oxygen

offspring cells Two genetically and morphologically similar products of equal cell division, or mitosis

organelle "Little organ"; a distinct intracellular structure composed of a complex of macromolecules and small molecules; for example, nuclei, mitochondria, and plastids

parasite Organism that lives on or in an organism of a different species and obtains nutrients from it

parent cell Cell that, by mitosis, gives rise to two or more nuclei or cells able to continue existing

phaeoplast Brown plastid, the membrane-bounded photosynthetic structure of brown algal cells

phagocytosis Ingestion, by a cell, of solid particles by flowing over and engulfing them whole

phloem A plant vascular tissue that transports nutrients; in trees, the inner bark

photosynthate Chemical products of photosynthesis

photosynthesis Production of organic compounds from carbon dioxide and water by using light energy captured by chlorophyll

phytoplankton Free-floating microscopic photosynthetic organisms

pinocytosis Ingestion, by a cell, of liquid droplets by engulfing them whole

plant Member of the kingdom Plantae, photoautotrophic sexual green eukaryotes that develop from embryos; they include mosses, ferns, conifers, and flowering plants

plastid Cytoplasmic, photosynthetic pigmented organelle (such as a chloroplast) or its nonphotosynthetic derivative

polysaccharide Molecule of indefinite size composed of a chain or network of sugar molecules; for example, starch and cellulose

porphyrin Nitrogen-containing heterocyclic organic compound; its derivatives include chlorophyll and heme

prokaryote Cell or organism composed of cells lacking a membrane-bounded nucleus, membrane-bounded organelles, and DNA coated with histone proteins

protein Molecule consisting of a long folded chain of linked amino acids; some are enzymes, which hasten chemical reactions in living organisms; others play a structural role (for example, tubulin, actin, and myosine)

protist Microscopic member of the kingdom Protoctista; an informal name for heterotrophic (protozoan) or autotrophic (alga) eukaryotic microorganism

protoctist Member of the kingdom Protoctista, eukaryotic heterotrophic and autotrophic microorganisms and their larger descendants, none of which form embryos; they include diatoms, dinoflagellates, brown seaweeds and other algae, ciliates, amoebas, malarial parasites, slime molds, slime nets, and many other groups

pyrenoid Proteinaceous structure inside some plastids; serves as a center of starch formation

respiration Oxidative breakdown of food molecules and release of energy from them; the terminal electron acceptor is inorganic and may be oxygen or, in anaerobic organisms, nitrate, sulfate, or nitrite

rhodoplast Red plastid, the membrane-bounded photosynthetic structure of red algal cells

ribosome A spherical organelle composed of protein and ribonucleic acid; the site of protein synthesis

RNA Ribonucleic acid; a molecule composed of a linear sequence of nucleotides; can store genetic information; a component of ribosomes, it takes part in protein synthesis

sexual reproduction Reproduction leading to individual offspring having more than one parent

skeleton Hardened biogenic scaffolding; structural material often composed of calcium carbonate, silica, or calcium phosphate

spore Small or microscopic propagative unit capable of development into a mature or active organism; often desiccation- and heat-resistant

steroid Member of a class of biogenic organic compounds composed of four carbon rings and attached chemical groups; the class includes many hormones, such as testosterone, estrogen, cholesterol, and cycloartanol

stromatolite Laminated carbonate or silicate rocks, organo-sedimentary structures produced by growth, metabolism, trapping, binding, and/or precipitating of sediment by communities of microorganisms, principally cyanobacteria

symbiont Member of a symbiosis

symbiosis Intimate and protracted association between two or more organisms of different species

thylakoid Photosynthetic membrane, stacked in layers inside chloroplasts and photosynthetic bacteria

tissue Aggregation of similar cells organized into a structural and functional unit; a component of organs

trachea Air-conducting tube; windpipe

trichocyst Organelle underlying the surface of many ciliates and some mastigotes; capable of sudden discharge to sting prey

tubulin Microtubule proteins; the proteins (alpha and beta tubulin) that constitute the walls of the microtubules of undulipodia, mitotic and meiotic spindles, and nerve cells

undulipodium A cilium or eukaryotic "flagellum"; used primarily for locomotion and feeding

xylem Plant vascular tissue through which most of the water and minerals are conducted from the root to other parts of the plant; constitutes the wood of trees and shrubs

zygospore Large multinucleated resistant structure (resting spore) that results from the fusion of two fungal threads of opposite mating types

zygote Diploid nucleus or cell produced by the fusion of two haploid cells and destined to develop into a new organism

Glossary Acknowledgment

We are grateful to W. H. Freeman and Company for permission to reprint definitions which originally appeared in *Five Kingdoms: An Illustrated Guide to the Phyla of Life on Earth* by Lynn Margulis and Karlene V. Schwartz.

Index